Springer Tracts in Natural Philosophy

Volume 24

W0071834

Edited by B. D. Coleman

Co-Editors:

S. S. Antman · R. Aris · L. Collatz · J. L. Ericksen
P. Germain · W. Noll · C. Truesdell

W. Strieder · R. Aris

Variational Methods
Applied to Problems of Diffusion
and Reaction

With 12 Figures

Springer-Verlag New York Heidelberg Berlin 1973

William Strieder

University of Notre Dame, Department of Chemical Engineering
Notre Dame, Indiana 46556 / U.S.A.

Rutherford Aris

University of Minnesota, Department of Chemical Engineering and Materials Science
Minneapolis, Minnesota 55455/U.S.A.

AMS Subject Classifications (1970): Primary 49H05, 76R99, 76P05, 60J70
Secondary 82A40, 60J65, 60J60

ISBN-13: 978-3-642-65626-2   e-ISBN-13: 978-3-642-65624-8
DOI: 10.1007/978-3-642-65624-8

W. Strieder · R. Aris

Variational Methods Applied to Problems of Diffusion and Reaction

With 12 Figures

Springer-Verlag Berlin Heidelberg New York 1973

William Strieder

University of Notre Dame, Department of Chemical Engineering
Notre Dame, Indiana 46556 / U.S.A.

Rutherford Aris

University of Minnesota, Department of Chemical Engineering and Materials Science
Minneapolis, Minnesota 55455/U.S.A.

AMS Subject Classifications (1970): Primary   49H05, 76R99, 76P05, 60J70
Secondary   82A40, 60J65, 60J60

ISBN-13: 978-3-642-65626-2   e-ISBN-13: 978-3-642-65624-8
DOI: 10.1007/978-3-642-65624-8

To

Stephen Prager

amicitiae et admirationis ergo

Preface

This monograph is an account of some problems involving diffusion or diffusion with simultaneous reaction that can be illuminated by the use of variational principles. It was written during a period that included sabbatical leaves of one of us (W.S.) at the University of Minnesota and the other (R.A.) at the University of Cambridge and we are grateful to the Petroleum Research Fund for helping to support the former and the Guggenheim Foundation for making possible the latter. We would also like to thank Stephen Prager for getting us together in the first place and for showing how interesting and useful these methods can be. We have also benefitted from correspondence with Dr. A. M. Arthurs of the University of York and from the counsel of Dr. B. D. Coleman the general editor of this series.

Table of Contents

Chapter 1

Introduction and Preliminaries

1.1. General Survey

The calculus of variations has been patient of many and varied interpretations and applications to physical problems. Indeed the notion of a principle of least activity has at times exercised an almost mystical fascination, as if it were the counterpart in natural philosophy of the dictum of Ockham in metaphysical. In mathematics the calculus of variations has a place as a discipline in itself with applications in the proving of existence theorems and the provision of estimates. In applied mathematics it serves to unify the basis of certain sets of equations and leads to numerical approximations to their solutions or bounds on certain important functionals. It is with the last aspect that this monograph is concerned. Attention is restricted to a class of problems involving diffusion and reaction, for, desirable as it might be to review the whole scope of variational principles in natural philosophy, the severely modest ambition of this monograph better meets the limitations of our ability and the compass of a tract.

The broader aspect of the subject and applications in other areas may be explored elsewhere in the literature, which is indeed vast. The calculus of variations is a standard element of all books of "Mathematical Methods in ..." of which pride of place must be given to Courant and Hilbert's classic (1937). There is a wealth of introductory texts among which it is invidious to make a selection, while for applications in various areas one may turn to Funk (1962), Lanczos (1966), Lauwerier (1966), Gould (1966), Young (1969), Biot (1970), to mention but a handful. The volume of papers given at a conference in Chicago in 1965 (Donnelly, Herman and Prigogine 1966) is interesting in bringing together several aspects of applied mathematics. The contribution of Finlayson and Scriven in this volume is a valuable reminder that the approximation scheme that arises from a variational formulation may really be no different than that which comes from a more direct approach to the basic equations.

The problems we shall discuss here are distinguished by having to do with diffusion or the combination of diffusion and reaction. In addition some of them are concerned with the complexities of random media. Thus Chapter 2 treats the problem of estimating the effective diffusion coefficient of a porous medium whose properties are only known in a statistical sense. In preparation for this the bulk of this preparatory chapter is devoted to the discussion of the statistical functions that are needed to describe the properties of the random media. The most primitive of these is simply the porosity or average value of the fractional void volume, and one of the first concerns in Chapter 2 is to see if a bound can be put on the effective diffusion coefficient when only the porosity is known. Naturally such bounds are comparatively crude and for closer values two- or even three-point correlations are needed. Diffusion in the Knudsen regime, when the mean free path is much longer than the pore diameter, requires a statistical description of the pores. De Marcus' integral formulation of the Knudsen flow problem is used in §§ 2.7–2.9. In the third chapter two diffusion limited reaction problems are considered. The first is that of precipitation at discrete points of a super-saturated solution when the rate of precipitation is limited by the diffusion of the solute molecules to these centers. The second problem concerns the quenching of excited molecules such as arises in connection with the phenomenon of luminescence.

The last chapter deals with some problems of heterogeneous catalysis where variational methods can play a useful role in estimating the effective reaction rate and in the analysis of experimental data.

Since the equations of diffusion are of the same form as those for heat conduction, many of these problems can be reinterpreted in thermal, rather than material, terms. In a similar way, many of the heat conduction problems can be rewritten in the context of diffusion as Biot (1970) points out. The electrical and magnetic properties of random media can also be estimated on the lines of Chapter 2 since they depend on the solution of Laplace's equation. We have not attempted to make all these translations, but merely make reference to papers where they may be found.

1.2. Phenomenological Descriptions of Diffusion and Reaction

From the molecular viewpoint the phenomena of diffusion and reaction arise from the random motion and formation of individual molecules, and hence demand the modes of description proper to statistical mechanics. We shall however only require the grosser description of continuum theory and take over Fick's law without inquiring into its molecular basis. Thus the function $c(x)$ defines the concentration of a solute in a region $\mathcal{R}$ if for any volume $\mathcal{V}$ within $\mathcal{R}$ the total amount of the solute within $\mathcal{V}$ is

$$\int_{\mathcal{V}} c(x)\, d^3x \qquad (1.2.1)$$

Similarly the flux vector $j(x)$ is a vector field defined in $\mathcal{R}$ such that across any surface $\mathcal{S}$ lying in $\mathcal{R}$ the total flux is

$$\int_{\mathcal{S}} j(x) \cdot n\, d^2x. \qquad (1.2.2)$$

The flux given by this integral has the same sense as the normal. If the concentration and flux fields are not constant then each must be made a function of time t as well as of position x. (See the notes on nomenclature in § 1.7). In the steady state the total amount of solute within any volume remains constant so that if $\mathcal{S}$ is the surface $\partial\mathcal{V}$ encompassing $\mathcal{V}$

$$\int_{\partial\mathcal{V}} j(x) \cdot n\, d^2x = \int_{\mathcal{V}} \nabla \cdot j(x)\, d^3x = 0. \qquad (1.2.3)$$

To obtain the second form of this integral Gauss' theorem has been used, as it will be repeatedly throughout this study. If $j(x)$ is continuously differentiable the validity of Equation (1.2.3) for an arbitrary volume implies that

$$\nabla \cdot j = 0 \qquad (1.2.4)$$

at each point.

Fick's law for the diffusion of a solute in a homogeneous phase is a constitutive relation between j and c, namely

$$j = -D\nabla c \qquad (1.2.5)$$

where D is the diffusion coefficient. Thus the equation governing a steady-state concentration distribution is

$$\nabla \cdot (D\nabla c) = 0, \qquad (1.2.6)$$

or if D is constant

$$\nabla^2 c = 0. \qquad (1.2.7)$$

In the simple case of a slab of homogeneous material whose faces, say $x = 0$ and $x = L$, are maintained at constant concentrations, c_0 and c_L, the flux is constant, normal to the two faces and of magnitude

$$J = D(c_0 - c_L)/L. \qquad (1.2.8)$$

This equation can be used to define an effective diffusion coefficient when the material between the two faces is not homogeneous. For if we can solve the diffusion equation for an arbitrarily complex region between two such planes and determine the magnitude of the flux J per unit *total* area of slab cross section, then an effective diffusion coefficient D_e can be defined by

$$D_e = LJ/(c_0 - c_L).$$ (1.2.9)

when c_0 and c_L are concentrations per unit *total volume of slab*. So long as the dimensions of the fine structure of the material are small in comparison with L this apparent diffusion coefficient will be independent of L and gives a useful overall description of diffusion in a complex medium.

The reaction rate will be regarded as a function of the concentration at the site of reaction. Thus if $\hat{r}(c)\,d^2x$ is the rate of disappearance of the solute by reaction at a surface element d^2x of a liquid-solid interface with normal n pointing away from the solid phase, steady-state balance in which the flux to the surface just balances the reaction rate is

$$- j \cdot n = \hat{r}(c),$$ (1.2.10)

or

$$D\nabla c \cdot n = D\frac{\partial c}{\partial n} = \hat{r}(c).$$ (1.2.11)

At an inert boundary the normal flux, and hence the normal gradient of concentration, is zero.

1.3. Correlation Functions for Random Suspensions

Several statistical descriptions for a random suspension are formulated in the next three sections. Probability functions are defined which will be used to calculate effective diffusion coefficients in a porous solid, effective reaction rate constants for a suspension of reactant molecules, and effectiveness factors for molecular sieve catalysts.

In this section certain correlation functions are introduced and discussed for any statistically homogeneous (statistical properties do not vary with position) random suspension. The correlation functions are then calculated for a model void-solid pore structure generated by randomly placed, freely overlapping solid spheres (Fig. 1.3.1).

Correlations in a Statistically Homogeneous Porous Medium
We introduce a volume average $\langle f \rangle$ of a function f defined by

$$\langle f \rangle = \frac{1}{V} \int_{\mathscr{V}} f(x)\,d^3x$$ (1.3.1)

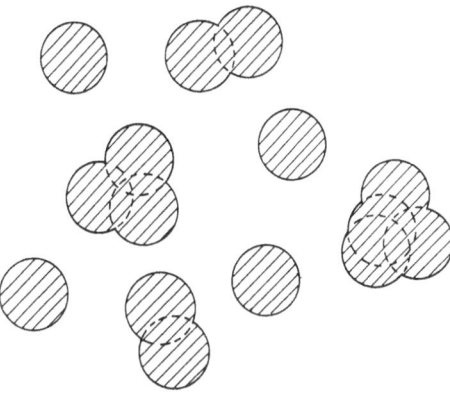

Fig. 1.3.1. Pore geometry generated by randomly overlapping spheres. Shaded areas denote the solid regions.

where the volume V covered by the integration over $\mathscr{V}$ is sufficiently large to extend over a representative sample of the random suspension. A complete knowledge of the function $g(x)$

$$g(x) = \begin{cases} 1 & x \text{ in the void} \\ 0 & x \text{ in the solid} \end{cases} \qquad (1.3.2)$$

would amount to a detailed specification of the void and solid regions of the porous medium. This is of course not obtainable in practice, but a partial description of the porous medium may be obtained by considering various volume averages of g (Debye, et al., 1949; 1957; Weissberg, 1963). Thus for example, the volume average $\langle g \rangle$ of g is just the void fraction Φ (ratio of void volume to total volume) for the porous medium.

Information about the shape of the void region may be obtained by averaging the product $g(x)\,g(x + \rho)$ with respect to x, keeping ρ fixed. The resulting two-point average

$$S(\rho) = \langle g(x)\,g(x + \rho) \rangle \qquad (1.3.3)$$

is, in spite of the discontinuous nature of g, a perfectly regular function of the relative position ρ of the two points concerned. The correlation $S(\rho)$ represents the probability that a line segment having the length and direction of the vector ρ will, when thrown at random into the porous material land with both ends in void regions. Measurements of

$S(\rho)$ can be made from x-ray scattering experiments (Debye and Bueche, 1949). In the absence of long range order the limits on S are

$$S(\rho) = \begin{cases} \Phi & \rho \to 0 \\ \Phi^2 & \rho \to \infty \end{cases} \tag{1.3.4}$$

The two-point correlation S in an isotropic suspension will depend on ρ, the magnitude of $\boldsymbol{\rho}$, alone.

To extend the statistical description further we define a three-point correlation $G(\boldsymbol{\rho}, \boldsymbol{\rho}')$, which gives the probability that three vertices of a triangle with the two sides $\boldsymbol{\rho}$ and $\boldsymbol{\rho}'$ thrown at random into the porous medium will all three lie in the void region.

$$G(\boldsymbol{\rho}, \boldsymbol{\rho}') = \langle g(x)\, g(x + \boldsymbol{\rho})\, g(x + \boldsymbol{\rho}')\rangle \tag{1.3.5}$$

Some generally valid limits of G are

$$G(\boldsymbol{\rho}, \boldsymbol{\rho}') = \begin{cases} \Phi & \rho, \rho' \to 0 \\ S(\boldsymbol{\rho}) & \rho' \to 0 \\ \Phi S(\boldsymbol{\rho}) & \rho' \to \infty \\ \Phi S(\boldsymbol{\rho}' - \boldsymbol{\rho}) & \rho, \rho' \to \infty, |\boldsymbol{\rho}' - \boldsymbol{\rho}| \text{ finite} \\ \Phi^3 & \rho, \rho', |\boldsymbol{\rho}' - \boldsymbol{\rho}| \to \infty \end{cases} \tag{1.3.6}$$

In an isotropic medium G will be a function of the distances ρ, ρ' and the angle between them.

It is possible to go beyond G to four-, five-, and six-point averages and so on. In an isotropic material G can still be obtained from a single cross-section since three points always lie in a plane, but this is no longer the case when more than three points are involved. It is desirable, therefore, that our calculations involve no averages of higher order than G. The problem for random suspensions will be to make optimal estimates of the quantity of interest from the partial information contained in G.

Void Fraction, Two- and Three-Point Correlations for a Bed of Overlapping Spheres

To determine these quantities, we must either make suitable measurements in the void-solid suspension, or we must know how the material was generated. A very simple method of generating a random suspension is one used by Weissberg (1963) who places Vn spheres of radius a in a volume V without any correlation between the position of the different spheres. (Fig. 1.3.1) The final result of this process will be a random suspension of spheres some isolated and some overlapping one another. If the solid region of a corresponding porous material consists of all points lying in the interior of one or more spheres, the random bed of solid spheres is a suitable model for the random-pore system of a porous medium.

To obtain an expression for the void fraction Φ in terms of the density of sphere centers n and the sphere radius a, we first calculate the probability of placing N_a centers at random in a finite volume V in such a way that a smaller volume v contains no centers. Since each random placement of a center is independent of the positions of the other centers, this probability is $\{(V - v)/V\}^{N_a} = \{1 - (nv/N_a)\}^{N_a}$. As N_a is made large, holding n and v fixed, the probability P_v that the volume v contains no centers becomes

$$P_v = \lim_{N_a \to \infty} (1 - nv/N_a)^{N_a} = e^{-nv} \tag{1.3.7}$$

Now Φ is just the probability that a randomly chosen point is not contained in any sphere, or that no sphere has its center within a distance a of the point. The probability the a region of volume $V_s = 4\pi a^3/3$ is empty of sphere centers is

$$\Phi = \exp(-nV_s) = \exp(-4\pi a^3 n/3) \tag{1.3.8}$$

The two-point correlation $S(\boldsymbol{\rho})$ defined as the probability that two points separated by a distance $\boldsymbol{\rho}$ both lie in the void region is also provided by equation (1.3.7). The probability that no sphere has its center within a region Ω consisting of all points lying within a distance a of either end of $\boldsymbol{\rho}$ is

$$S(\boldsymbol{\rho}) = \exp\left[-nV_\Omega(\rho)\right], \tag{1.3.9}$$

where V_Ω is the volume of region Ω. When ρ is larger than $2a$, the region Ω consists of two separate spheres each of radius a; for $\rho < 2a$, the two spheres overlap and V_Ω is correspondingly reduced, until for $\rho = 0$, it becomes just $4\pi a^3/3$

$$V_\Omega(\rho) = \begin{cases} \dfrac{4\pi a^3}{3}\left(1 + \dfrac{3\rho}{4a} - \dfrac{\rho^3}{16a^3}\right), & \rho < 2a \\[3mm] \dfrac{8\pi a^3}{3}, & \rho > 2a \end{cases} \tag{1.3.10}$$

Note that equations (1.3.8), (1.3.9), and (1.3.10) give the proper limits of $S(\boldsymbol{\rho})$ from equation (1.3.4).

The three-point correlation function $G(\boldsymbol{\rho}, \boldsymbol{\rho}')$ represents the probability that the three points on the vertices of a triangle of side $\boldsymbol{\rho}, \boldsymbol{\rho}'$, and $(\boldsymbol{\rho}' - \boldsymbol{\rho})$ all lie in the void region. The probability that spherical volumes of radius a centered at each of the three vertices are free of sphere centers is given in terms of $V_{\Omega'}(\boldsymbol{\rho}, \boldsymbol{\rho}')$, the volume of region Ω' of all points lying at a distance a or less from one or more of the three vertices of the triangle, by equation (1.3.7)

$$G(\boldsymbol{\rho}, \boldsymbol{\rho}') = \exp[-nV_{\Omega'}(\boldsymbol{\rho}, \boldsymbol{\rho}')] \qquad (1.3.11)$$

The evaluation of the volume $V_{\Omega'}$ is straightforward, the form of $V_{\Omega'}$ as presented by Weissberg and Prager (1962) is

$$V_{\Omega'}(\boldsymbol{\rho}, \boldsymbol{\rho}') \equiv \begin{cases} V_\Omega(\rho) + V_\Omega(\rho') + V_\Omega(\rho'') - 3V_s, & (l_1 > 2a) \\ V_\Omega(\rho) + V_\Omega(\rho') + V_\Omega(\rho'') - V_\Omega(l_1) - V_s, \\ \qquad (l_1 < 2a, l > a, l_1^2 > l_2^2 + l_3^2) \\ V_\Omega(\rho) + V_\Omega(\rho') + V_\Omega(\rho'') - 3V_s, \\ \qquad (l_1 < 2a, l > a, l_1^2 < l_2^2 + l_3^2) \\ V_s \sum_{i=1}^{3} F_G(\theta_i), & (l < a) \end{cases} \qquad (1.3.12)$$

where l_1, l_2, and l_3 are respectively, the largest, second largest, and smallest of the three lengths ρ, ρ', and $\rho''(=|\boldsymbol{\rho} - \boldsymbol{\rho}'|)$, the θ_i are the angles between the sides of the triangle, and

$$l = \rho\rho'\rho''/\{2(\rho^2\rho'^2 + \rho^2\rho''^2 + \rho'^2\rho''^2) - \rho^4 - \rho'^4 - \rho''^4\}^{1/2} \qquad (1.3.13)$$

is the radius of a circle passing through all three vertices. The function $F_G(\theta)$ is defined by

$$F_G(\theta) = \frac{1}{2\pi} \left\{ \frac{l}{a} \left(3 - \frac{l^2}{a^2}\sin^2\theta \right) \sin\theta \cot^{-1}\left[\frac{-l/a}{(1 - l^2/a^2)^{1/2}}\cos\theta \right] \right.$$
$$\left. + (l/a)^2(1 - l^2/a^2)^{1/2}\sin\theta\cos\theta + 2\tan^{-1}[(1 - l^2/a^2)^{1/2}\tan\theta] \right\}$$
$$(1.3.14)$$

The inverse tangent and cotangent appearing in equation (1.3.14) are to be taken between 0 and π.

1.4. Mean Free Path Statistics

Molecule-molecule collisions are neglected in the discussion of the so-called Knudsen diffusion (Strieder 1971). The diffusion of a molecule is a series of molecular free path jumps. A molecular free path begins with a molecule-pore wall collision, the molecule then traces out a straight line trajectory, and finally the free path ends with a subsequent molecule-pore wall collision. The statistical characterization of the porous medium for Knudsen diffusion reflects the need for information about these free paths.

We assume that the porous material is statistically homogeneous, i.e., that its properties do not vary *systematically* with position. We define the probability $h_\sigma(\boldsymbol{\rho}, \boldsymbol{n}, \boldsymbol{n}') d^3\boldsymbol{\rho}\, d^2\boldsymbol{n}\, d^2\boldsymbol{n}'$ that two points on the pore wall which can see one another have a relative position vector lying in a volume element $d^3\boldsymbol{\rho}$ about $\boldsymbol{\rho}$, and that the unit normal to the pore wall (taken to point into the void region) falls within the solid angle $d^2\boldsymbol{n}$ about $\boldsymbol{n}$ at the first point, and within $d^2\boldsymbol{n}'$ about $\boldsymbol{n}'$ at the second. In addition, we define the mean pore surface area σ which can be seen (i.e., reached by an unobstructed straight line) from a typical point on the void-solid surface, the pore wall surface s per unit total volume, an average pore diameter m_a which is four times the void volume to pore wall surface ratio, and the void fraction Φ (ratio of free void volume to total volume).

Both in real materials and in the overlapping sphere model it may happen that some of the voids are completely surrounded by solid material. Such "bubbles" of course do not contribute to mass transport, but they do contribute to the probability functions introduced above. Experimentally Φ can be obtained from the density of the porous material, and s has been measured by P. Debye, H. R. Anderson, Jr., and H. Brumberger using low angle scattering of x-rays (1957).

The probability functions, introduced for any statistically homogeneous void-solid distribution, will now be calculated for the void-solid structure generated when Vn spheres all of the same radius a are placed (Fig. 1.3.1) randomly into a large volume V without any correlation between the positions of different spheres. As in the previous section the solid region is taken to be all points which lie on the interior of one or more spheres; all those points lying on a sphere surface, but not in the interior of one or more overlapping spheres make up the pore wall surface.

We have already shown in section 1.3 that the probability P_v that a volume v is free of sphere centers is

$$P_v = \exp(-nv) \qquad (1.3.7 \text{ bis})$$

where n is the density of sphere centers. Also the void fraction Φ the probability that a randomly chosen point is not contained in any sphere was shown in section 1.3 to be

$$\Phi = \exp(-4\pi a^3 n/3) \qquad (1.3.8 \text{ bis})$$

The total surface area of the spheres is $4\pi a^2 n$ per unit volume, but only the fraction Φ of this is exposed, so that

$$s = 4\pi a^2 n \Phi \qquad (1.4.1)$$

The average pore diameter m_a (four times the void volume to pore wall surface ratio) for a bed of randomly overlapping solid spheres is

$$m_a^{-1} = \pi a^2 n \tag{1.4.2}$$

If we consider two points lying on the surfaces of different spheres, all values of $\boldsymbol{\rho}$, $\boldsymbol{n}$, and $\boldsymbol{n}'$ are equally likely, and the probability of falling within any specified intervals $d^3\boldsymbol{\rho}$, $d^2\boldsymbol{n}$, and $d^2\boldsymbol{n}'$ is just

$$\frac{d^3\boldsymbol{\rho}}{V} \cdot \frac{d^2\boldsymbol{n}}{4\pi} \cdot \frac{d^2\boldsymbol{n}'}{4\pi}$$

The probability $\mathscr{P}$ that two points are exposed and can see one another is zero unless $\boldsymbol{\rho} \cdot \boldsymbol{n} > 0$ and $\boldsymbol{\rho} \cdot \boldsymbol{n}' < 0$, since otherwise at least one point will be screened from the other by its own sphere. If these conditions are satisfied, then $\mathscr{P}$ is the probability that no third sphere has its center within a distance a of the line joining the two points, i.e., that no sphere center falls within the region formed by a cylinder of length ρ and radius a, capped at each end by a hemisphere, also of radius a,

$$\mathscr{P} = \exp\left[-\left(\frac{4\pi a^3}{3} + \pi a^2 \rho\right)n\right]$$
$$(\text{if} \quad \boldsymbol{\rho} \cdot \boldsymbol{n} \geq 0 \quad \text{and} \quad \boldsymbol{\rho} \cdot \boldsymbol{n}' \leq 0), \tag{1.4.3}$$
$$= 0 \quad (\text{otherwise})$$

Finally, the probability that two points on the surfaces of different spheres will be exposed and able to see one another, regardless of their relative positions and the orientation of the normals, is simply $\sigma \Phi^2 / sV$. In terms of these probabilities we have

$$h_\sigma \, d^3\boldsymbol{\rho} \, d^2\boldsymbol{n} \, d^2\boldsymbol{n}' = \frac{sV}{\sigma\Phi^2} \, \mathscr{P} \, \frac{d^3\boldsymbol{\rho}}{V} \frac{d^2\boldsymbol{n}}{4\pi} \frac{d^2\boldsymbol{n}'}{4\pi}$$

$$h_\sigma(\boldsymbol{\rho}, \boldsymbol{n}, \boldsymbol{n}') = \frac{s}{16\pi^2 \Phi \sigma} \exp\left(-\pi a^2 \rho n\right) \tag{1.4.4}$$

$$(\text{if} \quad \boldsymbol{\rho} \cdot \boldsymbol{n} \geq 0 \quad \text{and} \quad \boldsymbol{\rho} \cdot \boldsymbol{n}' \leq 0)$$
$$= 0 \quad (\text{otherwise})$$

Normalization of h_σ leads to

$$\sigma = 128\pi\Phi^2/s^2 \tag{1.4.5}$$

It is interesting to write the probability function h_σ in terms of the average pore diameter m_a, from equations (1.4.1), (1.4.2), (1.4.4), and (1.4.5) we have

$$h_\sigma(\boldsymbol{\rho}, \boldsymbol{n}, \boldsymbol{n}') = (32\pi^3 m_a^3)^{-1} \exp(-\rho m_a^{-1})$$

$$\text{(if } \boldsymbol{\rho} \cdot \boldsymbol{n} \geq 0 \quad \text{and} \quad \boldsymbol{\rho} \cdot \boldsymbol{n}' \leq 0) \qquad (1.4.6)$$

$$= 0 \quad \text{(otherwise)}$$

For the bed of spheres the probability that two points on the pore wall can see one another drops sharply and becomes quite small after a few pore diameters. Free paths much larger than m_a become unlikely due to the structure of the bed itself. If on the other hand the mean free path for molecule-molecule collision is much larger than m_a, clearly molecule collisions can be neglected.

1.5. Void Point—Surface Statistics

For those problems where a chemical reaction is catalyzed by an interface, it is helpful to develop a statistical description of an inhomogeneous medium which places emphasis on the interface surface. Such a description was introduced by Prager and Reck in a series of articles (Prager, 1963; Prager and Reck, 1965; Reck and Reck, 1968).

A total volume $\mathscr{V}$ contains a void region $\mathring{\mathscr{V}}$ through which some chemical species B^* diffuses, the remaining region $\mathscr{V} - \mathring{\mathscr{V}}$ does not contain the chemical species B^*, the species B^* reacts instantaneously and irreversibly upon contact with the interface $\partial\mathring{\mathscr{V}}$ between $\mathscr{V} - \mathring{\mathscr{V}}$ and the void region $\mathring{\mathscr{V}}$. For each point x in the void there is a minimum distance $\varepsilon(x)$ to the interface $\partial\mathring{\mathscr{V}}$; for points on the interface ε is of course zero (Fig. 1.5.1.).

If $\varepsilon(x)$ is the minimum distance from x to the interface $\partial\mathring{\mathscr{V}}$, then the normal vector $\boldsymbol{n}$ to that nearest surface element pointing into the void region $\mathring{\mathscr{V}}$ must be on a line drawn from the void point x to the nearest surface element. The gradient of any function is a vector whose magnitude is equal to the maximum change of that function and which points in the direction of that change, then we have

$$\nabla\varepsilon(x) = \boldsymbol{n} \qquad (1.5.1)$$

The suspension is characterized by specifying the probability $P(\varepsilon)\,d\varepsilon$ that a point chosen at random in $\mathscr{V}$ is in the void region $\mathring{\mathscr{V}}$, and lies at a distance between ε and $\varepsilon + d\varepsilon$ from the closest point on the interface $\partial\mathring{\mathscr{V}}$. If we consider the region $\mathscr{V} - \mathring{\mathscr{V}}$ to be composed of the interior points of randomly distributed, freely overlapping spheres all of radius a (Fig. 1.5.1), the probability of finding one sphere at a certain point is in no way affected by the presence of another sphere. The probability $P(\varepsilon)\,d\varepsilon$, that a randomly chosen point in the large volume $\mathscr{V}$ will lie in

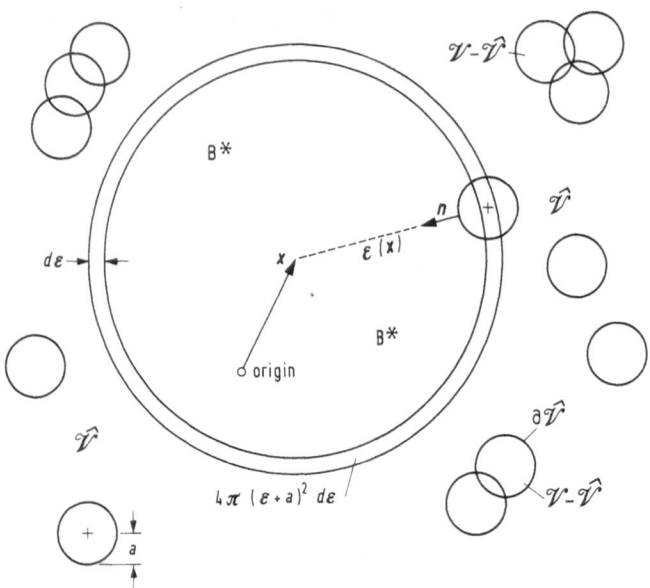

Fig. 1.5.1. Random spheres of radius a. The void region $\hat{\mathscr{V}}$, reaction zone $\mathscr{V} - \hat{\mathscr{V}}$, the interface $\partial\hat{\mathscr{V}}$, and the unit normal vector $\mathbf{n}$ to $\partial\hat{\mathscr{V}}$ are shown. The minimum distance from the point x in the void to the reactive interface is $\varepsilon(x)$. The regions about the point x of radius $\varepsilon + a$ from which spheres are excluded, and the adjacent shell of thickness $d\varepsilon$ in which at least one sphere center is found are also indicated.

the void $\hat{\mathscr{V}}$ with a nearest point on the interface $\partial\hat{\mathscr{V}}$ at a distance between ε and $\varepsilon + d\varepsilon$, is the product of the probability P_v of finding no sphere center within a spherical volume $4\pi(\varepsilon + a)^3/3$ multiplied by the probability P_s of at least one sphere center within the shell of radii $(\varepsilon + a)$ to $(\varepsilon + a) + d\varepsilon$

$$P(\varepsilon)\,d\varepsilon = P_s P_v \qquad (1.5.2)$$

or from (1.3.7)

$$P(\varepsilon)\,d\varepsilon = 4\pi(\varepsilon + a)^2\, n\, d\varepsilon \exp\left\{-4\pi(\varepsilon + a)^3\, n/3\right\} \qquad (1.5.3)$$

where n is the density of sphere centers.

1.6. Variational Principles Applied to the Diffusion Equation

We shall not attempt to set up a grand variational formulation from which all our cases can be deduced by various specializations, preferring to sketch the derivation of the equations rather lightly in the places where they arise, but there is a formulation of complementary variational principles which is perhaps less familiar and it will help to give an outline of it here. It follows the work of Noble (Noble, 1964; Noble and Sewell, 1972) and has been elaborated by Arthurs in numerous papers and in a monograph (Arthurs 1970). This monograph is a most valuable reference for this particular aspect of the calculus of variations, since it contains both the general formulation and a considerable number of applications. We shall follow Arthurs' presentation closely, specializing it to the operator which arises in the diffusion equations.

If q is any vector field, and u a scalar field defined on a region $\mathscr{V}$ with boundary $\partial \mathscr{V}$ then,

$$\int_{\mathscr{V}} q \cdot \nabla u \, d^3x = - \int_{\mathscr{V}} u \, \nabla \cdot q \, d^3x + \int_{\partial \mathscr{V}} u q \cdot n \, d^2x \qquad (1.6.1)$$

where n denotes the outward normal to $\partial \mathscr{V}$. The fields q and u are continuous with at least piecewise continuous derivatives; also $\mathscr{V}$ cannot have too exotic a shape, but in the present context this requirement is usually satisfied.

Consider the functional

$$\begin{aligned}
\mathscr{F}(q, u) &= \int_{\mathscr{V}} \{q \cdot \nabla u - \mathscr{H}(x, q, u)\} \, d^3x \\
&\quad - \int_{\partial \mathscr{V}} \{(q \cdot n) u - \mathscr{A}(q, u)\} \, d^2x \\
&= - \int_{\mathscr{V}} \{u \, \nabla \cdot q + \mathscr{H}(x, q, u)\} \, d^3x + \int_{\partial \mathscr{V}} \mathscr{A}(q, u) \, d^2x,
\end{aligned} \qquad (1.6.2)$$

and suppose that it has an extremum for $q = q_0$ and $u = u_0$. If we substitute

$$q = q_0 + \delta q_\delta, \qquad u = u_0 + \delta u_\delta \qquad (1.6.3)$$

into the functional we have

$$\mathscr{F}(q, u) = \mathscr{F}(q_0, u_0) + \delta \mathscr{F}_1 + \tfrac{1}{2} \delta^2 \mathscr{F}_2 + 0(\delta^3) \qquad (1.6.4)$$

where

$$\begin{aligned}
\mathscr{F}_1 &= \int_{\mathscr{V}} \{q_0 \cdot \nabla u_\delta + q_\delta \cdot \nabla u_0 - u_\delta \mathscr{H}_u - q_\delta \cdot \mathscr{H}_q\} \, d^3x \\
&\quad + \int_{\partial \mathscr{V}} \{q_\delta \cdot \mathscr{A}_q + u_\delta \mathscr{A}_u - (q_\delta \cdot n) u_0 - (q_0 \cdot n) u_\delta\} \, d^2x \\
&= - \int_{\mathscr{V}} \{u_0 \, \nabla \cdot q_\delta + u_\delta \, \nabla \cdot q_0 + u_\delta \mathscr{H}_u + q_\delta \cdot \mathscr{H}_q\} \, d^3x \\
&\quad + \int_{\partial \mathscr{V}} \{q_\delta \cdot \mathscr{A}_q + u_\delta \mathscr{A}_u\} \, d^2x
\end{aligned} \qquad (1.6.5)$$

In this equation $\mathscr{H}_u$ denotes $(\partial\mathscr{H}/\partial u)_{q_0,u_0}$ and $\mathscr{H}_q$ the vector $\{(\partial\mathscr{H}/\partial q_x, \partial\mathscr{H}/\partial q_y, \partial\mathscr{H}/\partial q_z)\}_{q_0,u_0}$ so that

$$q_\delta \cdot \mathscr{H}_q = \left\{ \sum_{j=x,y,z} q_{\delta j} \frac{\partial \mathscr{H}}{\partial q_j} \right\}_{q_0,u_0} \tag{1.6.6}$$

The second form of $\mathscr{F}_1$ comes from the application of Gauss' theorem (1.6.1) to the first form. The second variation $\mathscr{F}_2$ can also be written in two different forms, namely

$$\mathscr{F}_2 = \int_\mathscr{V} \{2q_\delta \cdot \nabla u_\delta - q_\delta \cdot \mathscr{H}_{qq} \cdot q_\delta - 2q_\delta \cdot \mathscr{H}_{qu} u_\delta - \mathscr{H}_{uu} u_\delta^2\} \, d^3x$$

$$- \int_{\partial\mathscr{V}} \{2(q_\delta \cdot n) u_\delta - q_\delta \cdot \mathscr{A}_{qq} \cdot q_\delta - 2q_\delta \cdot \mathscr{A}_{qu} u_\delta - \mathscr{A}_{uu} u_\delta^2\} \, d^2x$$
$$\tag{1.6.7}$$

$$= - \int_\mathscr{V} \{2u_\delta(\nabla \cdot q_\delta) + q_\delta \cdot \mathscr{H}_{qq} \cdot q_\delta + 2q_\delta \cdot \mathscr{H}_{qu} u_\delta + \mathscr{H}_{uu} u_\delta^2\} \, d^3x$$

$$+ \int_{\partial\mathscr{V}} \{q_\delta \cdot \mathscr{A}_{qq} \cdot q_\delta + 2q_\delta \cdot \mathscr{A}_{qu} u_\delta + \mathscr{A}_{uu} u_\delta^2\} \, d^2x \tag{1.6.8}$$

In these equations

$$q_\delta \cdot \mathscr{H}_{qq} \cdot q_\delta = \sum_{j,k=x,y,z} q_{\delta j} \left(\frac{\partial^2 \mathscr{H}}{\partial q_j \partial q_k} \right)_{q_0,u_0} q_{\delta k} \tag{1.6.9}$$

and

$$q_\delta \cdot \mathscr{H}_{qu} = \sum_{j=x,y,z} q_{\delta j} \left(\frac{\partial^2 \mathscr{H}}{\partial q_j \partial u} \right)_{q_0,u_0}$$

Now if $q = q_0$, $u = u_0$ gives an extremum value to the functional $\mathscr{F}(q, u)$, the first variation $\mathscr{F}_1$ must vanish for all variations q_δ and u_δ. Then, using equation (1.6.5), we can write

$$\int_\mathscr{V} \{q_\delta \cdot (\nabla u_0 - \mathscr{H}_q) - u_\delta(\nabla \cdot q_0 + \mathscr{H}_u)\} \, d^3x$$

$$+ \int_{\partial\mathscr{V}} \{q_\delta \cdot (\mathscr{A}_q - n u_0) + u_\delta \mathscr{A}_u\} \, d^2x = 0$$

Hence $\mathscr{F}(q, u)$ will be stationary at $q = q_0$, $u = u_0$ if q_0 and u_0 satisfy the canonical equations

$$\nabla u_0 = \mathscr{H}_q \quad \text{in } \mathscr{V} \tag{1.6.10}$$

$$\nabla \cdot q_0 = -\mathscr{H}_u \quad \text{in } \mathscr{V} \tag{1.6.11}$$

$$\mathscr{A}_u = 0, \quad \mathscr{A}_q - n u_0 = 0 \quad \text{on } \partial\mathscr{V} \tag{1.6.12}$$

There are two ways of putting trial functions into $\mathscr{F}(q, u)$ and it will be shown that they both provide a useful bound. In the first we choose a scalar field u and associate with it the vector field that satisfies

$$\nabla u = \mathscr{H}_q(q, u) \quad \text{in } \mathscr{V} \tag{1.6.13}$$

and

$$nu = \mathscr{A}_q(q, u) \quad \text{on } \partial\mathscr{V}$$

Let the solution of this equation for q be denoted by $Y(u)$ and substitute it in the first form of $\mathscr{F}$ given in equation (1.6.2), i.e., let

$$\mathscr{J}(u) = \mathscr{F}(Y(u), u) \tag{1.6.14}$$

Now if $u = u_0$, where u_0 satisfies equations (1.6.10) and (1.6.12) then $Y(u_0)$ will be q_0 and $\mathscr{F}$ will be stationary. In other words $\mathscr{J}(u_0)$ will be stationary at $u = u_0$ and

$$\mathscr{J}(u) = \mathscr{F}(q_0, u_0) + \tfrac{1}{2}\mathscr{J}_2 + 0(\delta^3) \tag{1.6.15}$$

Now from (1.6.10) and (1.6.13)

$$\nabla(u - u_0) = \delta\,\nabla u_\delta = \mathscr{H}_q(q, u) - \mathscr{H}_q(q_0, u_0)$$
$$= \mathscr{H}_{qu}(u - u_0) + \mathscr{H}_{qq} \cdot (q - q_0) + \dots$$

so that by substituting into (1.6.7) for $\delta q_\delta = q - q_0$ and $\delta u_\delta = u - u_0$ the term $\mathscr{J}_2$ can be written

$$\begin{aligned}
\mathscr{J}_2 = \int_{\mathscr{V}} &\{(Y(u) - q_0) \cdot \mathscr{H}_{qq} \cdot (Y(u) - q_0) \\
&- (u - u_0)\,\mathscr{H}_{uu}(u - u_0)\} \, d^3x \\
- \int_{\partial\mathscr{V}} &\{(Y(u) - q_0) \cdot \mathscr{A}_{qq} \cdot (Y(u) - q_0) \\
&- (u - u_0)\,\mathscr{A}_{uu}(u - u_0)\} \, d^2x
\end{aligned} \tag{1.6.16}$$

Thus $\mathscr{J}(u)$ is a minimum if $\mathscr{J}_2 \geq 0$ and a maximum if $\mathscr{J}_2 \leq 0$.

Alternatively we can choose a vector field q and associate with it a scalar field $\Theta(q)$ which is the solution of

$$\nabla \cdot q = -\mathscr{H}_u(q, u)$$
$$\mathscr{A}_u(q, u) = 0 \tag{1.6.17}$$

Then substituting in the second form of (1.6.2) we have

$$\mathscr{G}(q) = \mathscr{F}(q, \Theta(q)) = -\int_{\mathscr{V}} \{\Theta(q)\,\nabla \cdot q + \mathscr{H}(x, q, \Theta(q))\} \, d^3x$$
$$+ \int_{\partial\mathscr{V}} \mathscr{A}(q, \Theta(q)) \, d^2x \tag{1.6.18}$$

Again $\mathscr{G}(q)$ is stationary at $q = q_0$ for then $\Theta(q) = u_0$ and substituting in (1.6.8),

$$\mathscr{G}(q) = \mathscr{F}(q_0, u_0) + \tfrac{1}{2}\mathscr{G}_2 + 0(\delta^3) \qquad (1.6.19)$$

where

$$\mathscr{G}_2 = -\int_{\mathscr{V}} \{(q - q_0)\cdot\mathscr{H}_{qq}\cdot(q - q_0) - (\Theta(q) - u_0')\mathscr{H}_{uu}(\Theta(q) - u_0))\} d^3x$$

$$+ \int_{\partial\mathscr{V}} \{(q - q_0)\cdot\mathscr{A}_{qq}\cdot(q - q_0) - (\Theta(q) - u_0)\mathscr{A}_{uu}(\Theta(q) - u_0))\} d^2x$$

$$(1.6.20)$$

But we observe that $\mathscr{G}_2$ has a form similar to $-\mathscr{J}_2$ so that the conditions that make one positive will make the other negative. For example, if $\mathscr{G}_2 \leq 0$ then $\mathscr{G}(q)$ will have a maximum at $q = q_0$. Thus we have Noble's theorem (see Arthurs, 1970, p. 26).

The functionals $\mathscr{F}$, $\mathscr{J}$, and $\mathscr{G}$, defined by equations (1.6.2), (1.6.14) and (1.6.18) respectively, are stationary at $q = q_0, u = u_0$ when q and u satisfy equations (1.6.10)–(1.6.12). Moreover for small departures from q_0 and u_0

$$\mathscr{G}(q) \leq \mathscr{F}(q_0, u_0) \leq \mathscr{J}(u) \quad \text{if} \quad \mathscr{G}_2 \leq 0, \mathscr{J}_2 \geq 0 \qquad (1.6.21)$$

or

$$\mathscr{J}(u) \leq \mathscr{F}(q_0, u_0) \leq \mathscr{G}(q) \quad \text{if} \quad \mathscr{J}_2 \leq 0, \mathscr{G}_2 \geq 0 \qquad (1.6.22)$$

1.7. Notation

A few notes on notational conventions which we have tried to follow may be useful. Position relative to a fixed origin is denoted by the vector x, while ρ is used for relative position vectors. Only one integral sign is used for volume or surface integrals and the volume and surface elements are distinguished by the notations d^3x or d^2x.

When a functional is to be minimized or maximized the extremizing function may have a suffix 0 and the deviation from it a suffix 1. Thus if

$$\mathscr{J} = \int_{\mathscr{V}} \{(\nabla u)^2 + \gamma u^2\} d^3x$$

is being examined we may substitute $u = u_0 + u_1$ to give

$$\mathscr{J} = \mathscr{J}_0 + \mathscr{J}_1 + \mathscr{J}_2$$

where

$$\mathscr{I}_0 = \int_{\mathscr{V}} \{(\nabla u_0)^2 + \gamma u_0^2\} \, d^3x$$

$$\mathscr{I}_1 = 2 \int_{\mathscr{V}} \{(\nabla u_0) \cdot (\nabla u_1) + \gamma u_0 u_1\} \, d^3x$$

$$\mathscr{I}_2 = \int_{\mathscr{V}} \{(\nabla u_1)^2 + \gamma u_1^2\} \, d^3x.$$

The suffix 1 on the functional $\mathscr{I}_1$ will always denote the terms which are linear in u_1 and which of course must vanish for all u_1 if u_0 gives the extremum. The suffix 0 on the extremizing functions will be dropped once the equation for it is established and no confusion can arise. The exception to this convention is in the previous section where δu_δ has been used ephemerally in place of u_1.

Chapter 2

Diffusion Through a Porous Medium

2.1. Introduction

The principal object of this chapter is the estimation of the effective diffusion coefficient D_e of a solute in a gaseous or liquid mixture when the mixture is in the presence of a suspended solid phase. Bounds on D_e are calculated for an isotropic suspension whose only known statistical property is the void fraction, these are rigorously shown to be the "best possible bounds" that can be obtained with such limited information. Improved upper bounds on D_e are obtained when an explicit model pore structure is generated from randomly overlapping, solid spheres.

Diffusion in small pores is discussed in the final sections of the chapter; the so-called Knudsen diffusion occurs when the mean free path for molecule-molecule collisions within the pores is large compared with the average pore radius. An upper bound on the permeability of a porous medium to Knudsen diffusion is formulated, and calculations are given for Knudsen diffusion through a bed of randomly overlapping spheres.

2.2. Diffusion Through an Isotropic Porous Medium

When steady state diffusion of a solute occurs across a bed of suspended solid, one usually assumes that the mean concentration gradient $\boldsymbol{\alpha}$

$$\boldsymbol{\alpha} = \langle \nabla c \rangle \tag{2.2.1}$$

and the mean flux $\boldsymbol{J}$

$$\boldsymbol{J} = - \langle D(\boldsymbol{x}) \nabla c \rangle \tag{2.2.2}$$

are related by a measurable diffusion constant D_e through Fick's law

$$\boldsymbol{J} = - D_e \boldsymbol{\alpha} \tag{2.2.3}$$

The brackets refer to a volume average and are defined by

$$\langle \dots \rangle = \frac{1}{V} \int_{\mathscr{V}} \dots d^3 x \qquad (2.2.4)$$

where the volume of integration is sufficiently large to extend over a representative sample of the random suspension. To calculate J from (2.2.2) in terms of the local diffusion coefficient $D(x)$ we should determine the concentration $c(x)$ at every point x in the void of the porous medium. In concentrated suspensions of solid a complete solution of the diffusion equation is no longer possible, not only because of the mathematical difficulty, but also for the more basic reason that the solid-void interface, where the boundary conditions are given, is not known in detail, but only in a statistical sense. Since any statistical description involving a finite number of correlation functions will necessarily be incomplete an exact diffusion coefficient D_e is out of the question. Nevertheless, we may ask whether, even with a limited knowledge of the solid-void distribution, it is not at least possible to place some sort of bounds on D_e.

Rigorous bounds on D_e will be derived in the next four sections for an isotropic suspension whose only known statistical property is the void fraction. We shall show that any improvement in these bounds will require additional statistical information about the solid-void interface, and that improved bounds can be written in terms of certain two- and three-point correlations characterizing the geometry of the void-solid structure.

To describe the suspension we introduce the function $g(x)$ defined to be unity if the point x lies in the void region, and zero if x falls inside one of the suspended particles. In practice we are not given $g(x)$, but only some of the volume averages g generates. The void fraction Φ

$$\Phi = \langle g(x) \rangle, \qquad (2.2.5)$$

the two-point average $S(\rho)$

$$S(\rho) = \langle g(x) g(x + \rho) \rangle, \qquad (2.2.6)$$

and the three-point average $G(\rho, \rho')$

$$G(\rho, \rho') = \langle g(x) g(x + \rho) g(x + \rho') \rangle \qquad (2.2.7)$$

are discussed in section 1.3. Note the point already made in section 1.3 that it is not practical to go beyond $G(\rho, \rho')$ in the statistical characterization of an isotropic suspension. The bounds on D_e will be written in terms of these statistical functions.

To avoid difficulties arising from discontinuities in the concentration, we assume, to begin with, that the equilibrium distribution coefficient γ_e for the solute between the solid and void at the pore walls is unity.

Later we will discuss the simple extension of our results to the more common case when the solid is impenetrable. The diffusion coefficient has some value D_0 in the pores, and is zero in the solid

$$D(x) = D_0 g(x) \tag{2.2.8}$$

2.3. Variational Formulation for D_e

For steady state diffusion of a solute in such a suspension the trial concentration $u(x)$ that minimizes

$$\Sigma(u) = \langle D(x) \nabla u \cdot \nabla u \rangle \tag{2.3.1}$$

is the steady state concentration $c(x)$ in the volume. To demonstrate the minimum we give the trial function and the true concentration distribution the same fixed boundary values (not necessarily constant), at the entrances of the volume $\mathscr{V}$. Let u be written in the form $u_0 + u_1$, where u_1 is zero and $u_0 = c$ at the entrances of $\mathscr{V}$, so that

$$\Sigma(u) = \Sigma_0 + \Sigma_1 + \Sigma_2 \tag{2.3.2}$$

and Σ_1 contains the terms which are of the first order in u_1. Using the fact that the boundary values of u_1 are zero, this term can be written

$$\Sigma_1 = 2 \langle D \nabla u_1 \cdot \nabla u_0 \rangle \tag{2.3.3}$$
$$= 2 \langle \nabla \cdot (u_1 D \nabla u_0) \rangle - 2 \langle u_1 \nabla \cdot (D \nabla u_0) \rangle.$$

By using Gauss' theorem on the first term this part of Σ_1 can be written as an integral over $\mathscr{S}$, the void-solid interfacial area inside the slab. The element of this surface will be denoted by d^2x at the point x on the void-solid interface, and $n(x)$ is the unit normal to the interface pointing into the void. Thus

$$\Sigma_1 = 2V^{-1} \int_{\mathscr{S}} u_1 (-D \nabla u_0) \cdot n \, d^2x - 2 \langle u_1 \nabla \cdot (D \nabla u_0) \rangle \tag{2.3.4}$$

The variation u_1 is arbitrary on the surface $\mathscr{S}$ and within the volume $\mathscr{V}$, then Σ_1 vanishes if and only if the Euler-Lagrange equations

$$-D \nabla u_0 \cdot n = 0 \quad \text{on } \mathscr{S} \tag{2.3.5}$$

and

$$\nabla \cdot (D \nabla u_0) = 0 \quad \text{in } \mathscr{V}$$

are satisfied. We note first that $u_0 = c$ on the entrances to $\mathscr{V}$ and secondly that the trial function $u_0 (= c)$ that causes Σ_1 to vanish is the solution to the diffusion equation (1.2.6) within the suspension. Hence $\Sigma_1 = 0$

and $\Sigma(u) = \Sigma(c) + \Sigma_2$. But Σ is certainly never negative so that Σ_2 is positive unless u_1 is a constant, which by the boundary conditions can only be zero and the minimum value of Σ is established

$$\Sigma(c) \leq \Sigma(u) \qquad (2.3.6)$$

Strictly speaking the trial concentration u should be equal to the solute concentration at all the entrances to the volume $\mathscr{V}$ in order for inequality (2.3.6) to be valid. Note however that the presence of the V^{-1} factor (2.2.4) in the definition (2.3.1) of $\Sigma(u)$ in some cases allows us to drop these boundary conditions on u. For example in the case of a *very long* slab the contribution to $\Sigma(u)$ from the vicinity of the edges of the slab goes as L^{-1}, and for a long slab the values of u in this region do not effect $\Sigma(u)$. In the sections to follow, we will fix $\boldsymbol{\alpha} = \langle \nabla c \rangle = \langle \nabla u \rangle$, bound the values of $\boldsymbol{J}$, and thus obtain from (2.2.3) a bound on D_e.

To see how the D_e of (2.2.3) is related to $\Sigma(c)$ we break the trial function up into two parts, a mean concentration and a fluctuation $u'(x)$ about the mean, in terms of the gradients

$$\nabla u = \boldsymbol{\alpha} + \nabla u' \quad \text{where} \quad \langle \nabla u' \rangle = 0. \qquad (2.3.7)$$

Our object, as stated at the beginning of section 2.2 is to determine the mean flux $\boldsymbol{J}$. Fick's law (2.2.3), written in terms of an apparent diffusivity, implies that $\boldsymbol{J}$ does not depend on the details of the solute concentration on the outside boundaries of the suspension. However $\boldsymbol{J}$ does depend on the mean concentration gradient $\boldsymbol{\alpha}$ and we must consider the minimization of Σ subject to the restriction that

$$\boldsymbol{\alpha} = \langle \nabla c \rangle = \langle \nabla u \rangle \qquad (2.3.8)$$

is fixed.

The trial function (2.3.7) is substituted into (2.3.1) to obtain Σ in expanded form

$$\Sigma = D_0 \left[\Phi \alpha^2 + 2\boldsymbol{\alpha} \cdot \langle g(x) \nabla u' \rangle + \langle g(x) \nabla u' \cdot \nabla u' \rangle \right] \qquad (2.3.9)$$

Let $c'(x)$ be the fluctuation of the concentration field, i.e., the function such that $\nabla c = \boldsymbol{\alpha} + \nabla c'$, and consider a trial function of the form

$$u'(x) = \lambda c'(x) \qquad (2.3.10)$$

Here λ is an adjustable parameter and since the functional is minimized by c the value of λ will evidently be unity. Substitution of (2.3.10) into (2.3.9) together with the requirement that $\Sigma(\lambda)$ must have a minimum at $\lambda = 1$, leads to

$$-\boldsymbol{\alpha} \cdot \langle D \nabla c' \rangle = \langle D \nabla c' \cdot \nabla c' \rangle \qquad (2.3.11)$$

which can also be written

$$\Sigma(c) = -\boldsymbol{\alpha} \cdot \boldsymbol{J} \qquad (2.3.12)$$

On an intuitive basis the relation (2.3.12) might seem to be self-evident, since it merely expresses the rate of entropy production per unit volume in terms of mean rather than the local fluxes and gradients. Mathematically, however, equation (2.3.12) amounts to saying that

$$\langle D \, \nabla c \rangle \cdot \langle \nabla c \rangle = \langle D \, \nabla c \cdot \nabla c \rangle \qquad (2.3.13)$$

which is not at all true in general. That it is true for the actual concentration field requires the proof we have just given.

If a trial u' is used in place of the correct concentration distribution in (2.3.9) it will result in a value of Σ greater than the minimum value given by (2.3.12). Inequality (2.3.6) along with (2.2.3) and (2.3.12) becomes

$$D_e \alpha^2 \leq \Sigma(u) \qquad (2.3.14)$$

so that the calculation leads to an upper bound on D_e. If an arbitrary trial function $u'(x)$ has been chosen, we may consider others of the form $\lambda u'(x)$, and vary λ so as to minimize Σ. This time, however, the optimum λ is no longer unity, but is given by

$$\lambda = -\boldsymbol{\alpha} \cdot \langle D \, \nabla u' \rangle / \langle D \, \nabla u' \cdot \nabla u' \rangle \qquad (2.3.15)$$

and the upper bound for this λ is

$$D_e \leq D_0 \big[\Phi - (\boldsymbol{\alpha} \cdot \langle g \, \nabla u' \rangle)^2 \, (\alpha^2 \langle g \, \nabla u' \cdot \nabla u' \rangle)^{-1} \big] \qquad (2.3.16)$$

The last term on the right-hand side is always negative and an immediate, if somewhat trivial, consequence of (2.3.16) is the statement

$$D_e < D_0 \Phi \qquad (2.3.17)$$

2.4. Bounds on D_e for an Isotropic Suspension

To improve on the inequality (2.3.17) we must formulate a $u'(x)$ which is somehow related to the behavior of $g(x)$ in the neighborhood of the point x. For the average $\langle D \, \nabla u' \cdot \nabla u' \rangle$ in (2.3.16) to exist, u' must be smooth at all points, including points on the void-solid interface where g is discontinuous. The trial concentration fluctuation u' cannot therefore simply be a function of g. In addition to smoothness we have from (2.3.8) the requirement

$$\langle \nabla u' \rangle = 0. \qquad (2.4.1)$$

A suitable trial function satisfying both requirements is the weighted average of g,

$$u'(x) = \int h(\boldsymbol{\rho}) \big[g(x + \boldsymbol{\rho}) - \Phi \big] d^3\rho \qquad (2.4.2)$$

where $h(\rho)$ is the weighting function and the average extends over all space. Unlike g and u', $h(\rho)$ is not a stochastic function, and its precise dependence on ρ should, for best results, be chosen so as to minimize the right-hand side of (2.3.16).

With the trial function (2.4.2) the averages in the upper bound (2.3.16) become

$$\langle g(\boldsymbol{x})\,\nabla u'\rangle = \langle g(\boldsymbol{x}) \int h(\boldsymbol{\rho})\,\nabla[g(\boldsymbol{x} + \boldsymbol{\rho}) - \Phi]\,d^3\rho\rangle \qquad (2.4.3)$$

$$= \int h(\boldsymbol{\rho}) \frac{\partial S(\boldsymbol{\rho})}{\partial \boldsymbol{\rho}}\,d^3\rho, \qquad (2.4.4)$$

$$\langle g(\boldsymbol{x})\,\nabla u' \cdot \nabla u'\rangle = \iint h(\boldsymbol{\rho})\,h(\boldsymbol{\rho}') \frac{\partial}{\partial \boldsymbol{\rho}} \cdot \frac{\partial}{\partial \boldsymbol{\rho}'}\,G(\boldsymbol{\rho}, \boldsymbol{\rho}')\,d^3\rho\,d^3\rho' \quad (2.4.5)$$

and

$$\langle \nabla u' \cdot \nabla u'\rangle = \iint h(\boldsymbol{\rho})\,h(\boldsymbol{\rho}') \frac{\partial}{\partial \boldsymbol{\rho}} \cdot \frac{\partial}{\partial \boldsymbol{\rho}'}\,S(\boldsymbol{\rho}' - \boldsymbol{\rho})\,d^3\rho\,d^3\rho'. \quad (2.4.6)$$

In deriving (2.4.4) through (2.4.6) we have assumed the total volume V of the suspension is large.

We know that the concentration fluctuation u' must be a linear homogeneous function of the mean concentration gradient $\boldsymbol{\alpha}$, and it follows therefore that the dependence of h on $\boldsymbol{\alpha}$ must also be of this nature. Furthermore, since we are discussing an isotropic suspension, the dependence of h on $\boldsymbol{\rho}$ and $\boldsymbol{\alpha}$ must be invariant under rotation of the coordinate frame. The most general scalar function of $\boldsymbol{\rho}$ and $\boldsymbol{\alpha}$ satisfying these requirements is

$$h(\boldsymbol{\rho}) = (\boldsymbol{\alpha} \cdot \boldsymbol{\rho})\,H(\rho) \qquad (2.4.7)$$

where $H(\rho)$ is a function only of ρ, the magnitude of $\boldsymbol{\rho}$. If $h(\boldsymbol{\rho})$ has the form (2.4.7) the vectors $\langle g\,\nabla u'\rangle$ and $\boldsymbol{\alpha}$ are collinear as is to be expected for an isotropic system, and we can write

$$(\boldsymbol{\alpha} \cdot \langle g\,\nabla u'\rangle)^2 = \alpha^2 \langle g\,\nabla u'\rangle \cdot \langle g\,\nabla u'\rangle. \qquad (2.4.8)$$

We now turn our attention back to inequality (2.3.16) and use (2.4.8) to obtain

$$D_e \leq D_0 \big[\Phi - \langle g\,\nabla u'\rangle \cdot \langle g\,\nabla u'\rangle\,(\langle g\,\nabla u' \cdot \nabla u'\rangle)^{-1}\big] \qquad (2.4.9)$$

Note that direct substitution of equations (2.4.4) and (2.4.5) provides an upper bound in terms of the three-point average $G(\boldsymbol{\rho}, \boldsymbol{\rho}')$. If we want a bound on D_e for any isotropic suspension where the only known statistical property is the void fraction, we must assume that $G(\boldsymbol{\rho}, \boldsymbol{\rho}')$ is not known. The three-point correlation $G(\boldsymbol{\rho}, \boldsymbol{\rho}')$ can be eliminated from inequality (2.4.9) by the use of the Schwartz inequality

$$\langle (1 - g) \rangle \langle (1 - g) \nabla u' \cdot \nabla u' \rangle \geq \langle (1 - g) \nabla u' \rangle \cdot \langle (1 - g) \nabla u' \rangle, \quad (2.4.10)$$

or with (2.2.5) and (2.4.1)

$$(1 - \Phi) \langle (1 - g) \nabla u' \cdot \nabla u' \rangle \geq \langle g \nabla u' \rangle \cdot \langle g \nabla u' \rangle. \quad (2.4.11)$$

Applying (2.4.11) to (2.4.9) we obtain the weaker inequality

$$D_e < D_0 \left[\Phi - \frac{(1 - \Phi) \langle g \nabla u' \rangle \cdot \langle g \nabla u' \rangle}{(1 - \Phi) \langle \nabla u' \cdot \nabla u' \rangle - \langle g \nabla u' \rangle \cdot \langle g \nabla u' \rangle} \right] \quad (2.4.12)$$

but it is one from which explicit bounds on D_e can be calculated. There is no simple way to measure $G(\boldsymbol{\rho}, \boldsymbol{\rho}')$ it is known only for some model pore structures such as a random bed of solid spheres (see sec. 1.3), we shall return to a discussion of $G(\boldsymbol{\rho}, \boldsymbol{\rho}')$ later in the next section.

The upper bound on D_e after substitution of (2.4.4) and (2.4.6) becomes

$$D_e/D_0 < \Phi - (1 - \Phi) \left[(1 - \Phi) K_\Phi - 1 \right]^{-1} \quad (2.4.13)$$

where

$$K_\Phi = \frac{\displaystyle\iint h(\boldsymbol{\rho}) \, h(\boldsymbol{\rho}') \frac{\partial}{\partial \boldsymbol{\rho}} \cdot \frac{\partial}{\partial \boldsymbol{\rho}'} S(\boldsymbol{\rho}' - \boldsymbol{\rho}) \, d^3\boldsymbol{\rho} \, d^3\boldsymbol{\rho}'}{\left[\displaystyle\int h(\boldsymbol{\rho}) \frac{\partial S(\boldsymbol{\rho})}{\partial \boldsymbol{\rho}} d^3\boldsymbol{\rho} \right] \cdot \left[\displaystyle\int h(\boldsymbol{\rho}') \frac{\partial S(\boldsymbol{\rho}')}{\partial \boldsymbol{\rho}'} d^3\boldsymbol{\rho}' \right]} \quad (2.4.14)$$

To minimize the upper bound we must find the $h(\boldsymbol{\rho})$ that minimizes K_Φ; $h(\boldsymbol{\rho})$ should satisfy the Euler-Lagrange relation (see Appendix 2.4.A)

$$\int h(\boldsymbol{\rho}') \frac{\partial}{\partial \boldsymbol{\rho}} \cdot \frac{\partial}{\partial \boldsymbol{\rho}'} S(\boldsymbol{\rho}' - \boldsymbol{\rho}) \, d^3\boldsymbol{\rho}' = \boldsymbol{\alpha} \cdot \frac{\partial}{\partial \boldsymbol{\rho}} S(\boldsymbol{\rho}) \quad (2.4.15)$$

Equation (2.4.15) is readily solved, for example by the use of Fourier transforms, to give

$$h(\boldsymbol{\rho}) = -(\boldsymbol{\alpha} \cdot \boldsymbol{\rho})/4\pi\rho^3 \quad (2.4.16)$$

For this optimum $h(\boldsymbol{\rho})$ and the general boundary conditions (1.3.4) of $S(\boldsymbol{\rho})$, the minimum value of K_Φ is

$$K_\Phi = 3 \left[\Phi(1 - \Phi) \right]^{-1}.$$

The inequality (2.4.13) becomes after some reduction

$$D_e/D_0 < \Phi \left[1 + \tfrac{1}{2}(1 - \Phi) \right]^{-1} \quad (2.4.17)$$

Like (2.3.17), over which it is a clear improvement, the inequality (2.4.17) involves only the volume fraction, and does not depend on the correlations $S(\rho)$ or $G(\rho, \rho')$. It is a very general result valid for suspended particles of any shape at any value of Φ; the only restriction is that the suspension be isotropic.

Before ending this section, we must recall that all our calculations so far have been based on a value of 1 for the equilibrium distribution coefficient γ_e, whereas in most suspensions the solute does not penetrate into the solid phase at all, corresponding to $\gamma_e = 0$. Since the diffusion coefficient in the solid is zero anyway, this does not affect the concentration distribution in the void region, and the mean flux is the same whether or not the solute can enter the suspended particles. The mean concentration gradient is reduced by a factor Φ in going from $\gamma_e = 1$ to $\gamma_e = 0$, and the diffusion coefficient D_e' in a suspension of impenetrable particles is therefore

$$D_e' = \frac{D_e}{\Phi} < D_0 \left[1 + \frac{1}{2}(1 - \Phi) \right]^{-1} \qquad (2.4.18)$$

Appendix 2.4.A. Euler-Lagrange Equations for (2.4.13)

In some cases, a trial function is written in terms of some unknown function, but in a form much simpler than the complete solution of the problem [e.g. (2.4.2), (3.3.16) and (4.7.5) of this text]. An example is provided by the variational upper bound (2.4.13) expressed in terms of the unknown function $h(\rho)$. We will select the form of $h(\rho)$ that minimizes the variational upper bound, such extremizing functions are solutions to the Euler-Lagrange equations.

In fact, the minimum of the upper bound (2.4.13) is obtained for that function $h(\rho)$ that minimizes

$$K_\Phi(h) = \frac{\displaystyle\iint h(\rho)\, h(\rho')\, \frac{\partial}{\partial \rho} \cdot \frac{\partial}{\partial \rho'}\, S(\rho' - \rho)\, d^3\rho\, d^3\rho'}{\left(\displaystyle\int h(\rho)\, \frac{\partial S(\rho)}{\partial \rho}\, d^3\rho \right) \cdot \left(\displaystyle\int h(\rho')\, \frac{\partial S(\rho')}{\partial \rho'}\, d^3\rho' \right)} \qquad (2.4.14)$$

where the integration extends over all values of ρ. Because the suspension is isotropic, the S-correlation depends only on the magnitude of its argument. A second consequence of isotropy already discussed is the collinearity of the vectors α and $\int h(\rho)\,[\partial S(\rho)/\partial \rho]\, d^3\rho$ which results in the equation

$$\alpha^2 \left(\int h(\rho)\, \frac{\partial S(\rho)}{\partial \rho}\, d^3\rho \right)^2 = \left(\alpha \cdot \int h(\rho)\, \frac{\partial S(\rho)}{\partial \rho}\, d^3\rho \right)^2 \qquad (2.4.8)$$

The functional $K_\phi(h)$ becomes

$$K_\phi(h) = \frac{\alpha^2 \iint h(\boldsymbol{\rho})\, h(\boldsymbol{\rho}') \frac{\partial}{\partial \boldsymbol{\rho}} \cdot \frac{\partial}{\partial \boldsymbol{\rho}'} S(\boldsymbol{\rho}' - \boldsymbol{\rho})\, d^3\rho\, d^3\rho'}{\left(\boldsymbol{\alpha} \cdot \int h(\boldsymbol{\rho}) \frac{\partial S(\boldsymbol{\rho})}{\partial \boldsymbol{\rho}}\, d^3\rho \right)^2} \qquad (2.4.19)$$

We see from the form (2.4.6) that K_ϕ is positive and must possess a minimum. To find this minimum, we define the trial function $h = h_0 + h_1$; note that it is more convenient to work with

$$\tilde{h}_i(\boldsymbol{\rho}) = \frac{\alpha^2 h_i(\boldsymbol{\rho})}{\left(\boldsymbol{\alpha} \cdot \int h_0(\boldsymbol{\rho}) \frac{\partial S(\boldsymbol{\rho})}{\partial \boldsymbol{\rho}}\, d^3\rho \right) K_\phi(h_0)}, \qquad i = 0, 1 \quad (2.4.20)$$

$$K_\phi(h) = K_\phi(\tilde{h}),$$

and write the term first order in the variation of h (or $\tilde{h}$)

$$\frac{K_{\phi 1}}{K_{\phi 0}^2} = \frac{2}{\alpha^2} \int \tilde{h}_1(\boldsymbol{\rho}) \left[\int \tilde{h}_0(\boldsymbol{\rho}') \frac{\partial}{\partial \boldsymbol{\rho}} \cdot \frac{\partial}{\partial \boldsymbol{\rho}'} S(\boldsymbol{\rho}' - \boldsymbol{\rho})\, d^3\rho' - \boldsymbol{\alpha} \cdot \frac{\partial S(\boldsymbol{\rho})}{\partial \boldsymbol{\rho}} \right] d^3\rho$$

$$(2.4.21)$$

where we have used the property $S(|\boldsymbol{\rho}' - \boldsymbol{\rho}|)$ for an isotropic suspension. The variation h_1 about h_0 is arbitrary, hence the function which satisfies the Euler-Lagrange equation

$$\int h(\boldsymbol{\rho}') \frac{\partial}{\partial \boldsymbol{\rho}} \cdot \frac{\partial}{\partial \boldsymbol{\rho}'} S(\boldsymbol{\rho}' - \boldsymbol{\rho})\, d^3\rho' - \boldsymbol{\alpha} \cdot \frac{\partial S(\boldsymbol{\rho})}{\partial \boldsymbol{\rho}} = 0 \qquad (2.4.22)$$

will give a minimum value to the functional $K_\phi(h)$. Note that the tilde and zero subscript are no longer necessary and have been dropped from the notation.

For the function $h(\boldsymbol{\rho})$ which satisfies (2.4.22), K_ϕ can be written

$$K_\phi(h) = \left[\boldsymbol{\alpha} \cdot \int h(\boldsymbol{\rho}) \frac{\partial S(\boldsymbol{\rho})}{\partial \boldsymbol{\rho}}\, d^3\rho \right]^{-1} \qquad (2.4.23)$$

2.5. Best Possible Bounds on D_e

The upper bound (2.4.17) on D_e derived in the last section is valid for any isotropic suspension, and the only statistical property that need be known is the void fraction Φ. We must now pose the question, is (2.4.17) the lowest possible upper bound on D_e for an isotropic suspension of solid whose only known property is the void fraction? If an isotropic suspension can be constructed whose effective diffusion coefficient is equal to the upper bound (2.4.17), then we have calculated the best possible bound for our model. We shall show that (2.4.17) and zero are the best upper and lower bounds, respectively, and that any improvement will require additional statistical information about the structure of the suspension.

Consider a homogeneous slab through which a solute diffuses with a diffusion coefficient D_r. On the external surfaces of the slab $x = 0$ and L, the concentration is specified at the uniform values of c_0 and c_L respectively, and this creates within the slab a homogeneous concentration gradient, $\alpha = (c_L - c_0)/L$. A sphere of radius r_b is now replaced by a composite sphere (Fig. 2.5.1) consisting of an inner part of radius r_a in which the diffusion coefficient is D_a, and a concentric shell where the diffusion coefficient is D_b. Under what conditions will there be no change in the dissipation of the body?

If the concentration field outside of r_b remains unchanged there will be no change in the dissipation due to the displacement. To see this, we need only write the dissipation integral Σ_b for the sphere of radius r_b

$$\Sigma_b = \int_{\mathscr{V}_b} D(\boldsymbol{x}) \, \nabla c \cdot \nabla c \, d^3\boldsymbol{x} \qquad (2.5.1)$$

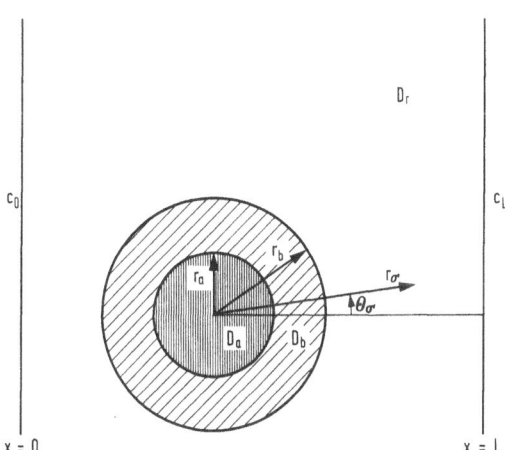

Fig. 2.5.1. Homogeneous slab and composite sphere. The composite sphere consists of an inner sphere with radius r_a and diffusion coefficient D_a, the concentric outer spherical shell has an outer radius r_b with a diffusion coefficient D_b, the homogeneous slab has diffusion coefficient D_r. Concentration is specified at the uniform values c_0 and c_L on the edges of the slab.

in terms of the surface integral from Gauss' theorem

$$\Sigma_b = \int_{\partial \mathscr{V}_b} cD_b \nabla c \cdot \boldsymbol{n} \, d^2\boldsymbol{x} \tag{2.5.2}$$

where $\boldsymbol{n}$ is an outward normal at the external surface $\partial \mathscr{V}_b$ of the sphere $\mathscr{V}_b$ of radius r_b. The concentration ($\gamma_e = 1$) and the normal component of the flux in the surface integral (2.5.2) are continuous across the interface $\partial \mathscr{V}_b$, and will *not* change with the addition of the composite sphere so long as the concentration remains the same outside of r_b. If such a system can be constructed, the volume average of the dissipation will be unchanged.

$$\Sigma = \langle D \, \nabla c \cdot \nabla c \rangle = D_r \alpha^2 \tag{2.5.3}$$

upon the substitution of one *or more* spheres. We note also from (2.3.12) and (2.2.3) that for a void-solid system

$$\Sigma = D_e \alpha^2 = D_r \alpha^2 \tag{2.5.4}$$

or

$$D_r = D_e.$$

Finally we will find that the existence of such a solution imposes a value on D_r.

In a spherical coordinate system $(r_\sigma, \theta_\sigma, \phi_\sigma)$ whose axis $\theta_\sigma = 0$ coincides (Fig. 2.5.1) with the flux field (x-axis) and whose center coincides with the center of the sphere considered, the concentration before the change is

$$c = c_p + \alpha r_\sigma \cos \theta_\sigma \tag{2.5.5a}$$

where c_p is the concentration at the origin, and $\alpha = (c_L - c_0)/L$ is defined by (2.3.8). This is therefore taken to be the concentration for

$$r_\sigma \geq r_b \tag{2.5.5b}$$

after the change. From Laplace's equation

$$c = c_p + (\alpha_b r_\sigma + \varepsilon_b r_\sigma^{-2}) \cos \theta_\sigma, \qquad r_a \leq r_\sigma \leq r_b, \tag{2.5.6}$$

and

$$c = c_p + \alpha_a r_\sigma \cos \theta_\sigma, \qquad r_\sigma \leq r_a. \tag{2.5.7}$$

The boundary conditions at r_a and r_b require that

$$\alpha r_b = \alpha_b r_b + \varepsilon_b r_b^{-2} \tag{2.5.8}$$

$$D_r \alpha = D_b(\alpha_b - 2\varepsilon_b r_b^{-3}) \tag{2.5.9}$$

$$\alpha_a r_a = \alpha_b r_a + \varepsilon_b r_a^{-2} \tag{2.5.10}$$

$$D_a \alpha_a = D_b(\alpha_b - 2\varepsilon_b r_a^{-3}) \tag{2.5.11}$$

In order that (2.5.8) through (2.5.11) be self-consistent, the determinant of the coefficients of α, α_b, ε_b, and α_a must vanish

$$\begin{vmatrix} 1 & -1 & -r_b^{-3} & 0 \\ D_r & -D_b & 2D_b r_b^{-3} & 0 \\ 0 & 1 & r_a^{-3} & -1 \\ 0 & D_b & -2D_b r_a^{-3} & -D_a \end{vmatrix} = 0, \qquad (2.5.12)$$

which reduces to

$$D_r = D_b + \cfrac{(r_a/r_b)^3}{\cfrac{1}{D_a - D_b} + \cfrac{1 - (r_a/r_b)^3}{3D_b}} \qquad (2.5.13)$$

When the above condition (2.5.13) is satisfied, the total dissipation within the slab is not influenced by the change made. Thus if this change is repeated indefinitely until all the original material is replaced by composite spheres and these are allowed to diminish to infinitesimal size, keeping the same value of (r_a/r_b), a macroscopically isotropic and homogeneous material can be constructed whose effective diffusion coefficient is exactly that given by (2.5.13).

Now suppose the inner spheres (Fig. 2.5.1) of radius r_a are penetrable* solid ($D_a = 0$) while the outer spheres are the void region ($D_b = D_0$). The void fraction is given by

$$\Phi = 1 - (r_a/r_b)^3 \qquad (2.5.14)$$

and the effective diffusion coefficient (2.5.13) is

$$D_e/D_0 = D_r/D_0 = \Phi \left[1 + \frac{1}{2}(1 - \Phi) \right]^{-1}, \qquad (2.5.15)$$

exactly the same form as the upper bound (2.4.17) on D_e/D_0. On the other hand suppose the inner sphere is the void region ($D_a = D_0$) and the outer sphere is the solid ($D_b = 0$) from (2.5.13) we have

$$D_e/D_0 = D_r/D_0 = 0 \qquad (2.5.16)$$

We know that D_e must be positive, then the lower bound on D_e for an isotropic suspension whose only known statistical property is the void fraction cannot be larger than zero.

---

* As explained at the end of section 2.2, we will assume the equilibrium distribution coefficient γ_e for the solute between the solid and void phases to unity $\gamma_e = 1$ and will relate the results to the impenetrable solid by (2.4.18).

It follows that (2.5.15) and (2.5.16) are the "best possible bounds" for the effective diffusion coefficient for a macroscopically homogeneous and isotropic suspension of solid which can be derived in terms of a void fraction and phase diffusion coefficients. To improve the bounds on D_e additional information on the statistics of the spatial distribution of the void and solid is needed.

An improved upper bound (Prager 1963) on D_e is readily obtained when we return to inequality (2.4.9) and substitute directly the equations (2.4.4) and (2.4.5) without using the Schwartz inequality (2.4.10)

$$\frac{D_e}{D_0} < \left\{ \Phi - \frac{\int h(\rho) \frac{\partial S(\rho)}{\partial \rho} d^3\rho \cdot \int h(\rho') \frac{\partial S(\rho')}{\partial \rho'} d^3\rho'}{\int\int h(\rho)\, h(\rho') \frac{\partial}{\partial \rho} \cdot \frac{\partial}{\partial \rho'} G(\rho, \rho')\, d^3\rho\, d^3\rho'} \right\} \quad (2.5.17)$$

While (2.5.17) represents an improvement over (2.4.17), its calculation will require a knowledge of the two-point $S(\rho)$ and three-point correlation $G(\rho, \rho')$ in addition to the void fraction Φ. These correlations are discussed in section (1.3), while $S(\rho)$ can be obtained experimentally there is no simple method of measuring $G(\rho, \rho')$. The three-point correlation can be obtained for model pore structure, for example, in a random bed of solid spheres $G(\rho, \rho')$ is given by Eq. (1.3.11).

2.6. Bounds on D_e for a Random Bed of Spheres

In the preceding sections an upper bound on the effective diffusion coefficient D_e as formulated in (2.3.14) and applied in (2.4.17) to a very general class of isotropic media. In most cases the statistical data required for an unrestricted calculation of these bounds is not available. One way to create the necessary statistics is to generate the pore structure in a specific manner. In this section variational upper bounds on D_e are obtained for a bed of spherical particles whose centers are randomly placed without restricting the spheres to nonoverlapping locations. The following calculation is due to Weissberg (1963).

If D_0 is the diffusion coefficient in the void space, and $g(x)$ is a function of the position vector x defined to have a value zero in the solid regions and unity in the void, the results of earlier sections of this chapter can be combined to give an upper bound on D_e/D_0 in terms of a trial concentration $u(x)$

$$D_e/D_0 < \langle g\, \nabla u \cdot \nabla u \rangle / \langle \nabla u \rangle^2 \quad (2.6.1)$$

The inequality (2.6.1) is a direct consequence of equations (2.2.8), (2.3.1),

(2.3.8), and (2.3.14). (Refer to footnote on page 29.) We must now select a proper trial concentration gradient ∇u and discuss the method of generating the void-solid structure.

One very simple model for the void-solid structure is obtained when solid spheres, all of the same radius a, are distributed at random regardless of their interpenetration. Thus the center of each sphere is located at coordinates chosen by some random process, and the spheres are allowed to overlap one another without restriction. A point lying on the interior of one or more spheres is a point in the solid, and a point that is not on the interior of any sphere is a point in the void (Fig. 1.3.1). In section 1.3 we discussed the probability (1.3.8) that a randomly chosen point is not contained in any sphere, and we showed this probability (the void fraction Φ) for the random bed of spheres could be written

$$\langle g \rangle = \Phi = \exp\{-4\pi a^3 n/3\}, \tag{2.6.2}$$

where n is the average number of sphere centers per unit volume.

The trial concentration gradient ∇u involves a sum of contributions from each sphere center, the center of sphere i being located at x_i. To this sum is added a constant vector A in the direction of the average concentration gradient, $\alpha = \langle \nabla c \rangle$

$$\nabla u(x) = A + m \sum_i \nabla F_a(\rho_i) \tag{2.6.3a}$$

where $\rho_i = x_i - x$ and, for spheres of radius a

$$F_a(\rho_i) = \begin{cases} -A \cdot \rho_i/\rho_i^3 & \rho_i > a \\ -A \cdot \rho_i/a^3 & \rho_i < a \end{cases} \tag{2.6.3b}$$

This form of the contribution from each center is suggested by the solution of the diffusion problem in which the obstruction is a single isolated sphere; its use here insures that the upper bound on D_e becomes exact as Φ approaches unity. The solutions for an isolated solid spherical obstruction in an infinite homogeneous diffusing medium are similar to those discussed in section 2.5. The scalar multiplier m in the trial concentration gradient (2.6.3a) is to be adjusted so as to minimize the right-hand side of inequality (2.6.1) when substitution is made for ∇u from (2.6.3).

When the trial concentration gradient (2.6.3a) is substituted into the upper bound (2.6.1) several summations result, using (2.6.2) we have

$$\frac{D_e}{D_0} < \frac{\Phi A^2 + 2mA \cdot \langle g \sum_i \nabla F_a(\rho_i) \rangle + m^2 \langle g [\sum_i \nabla F_a(\rho_i)]^2 \rangle}{[A + m\langle \sum_i \nabla F_a(\rho_i) \rangle]^2} \tag{2.6.4}$$

It is in the evaluation of these sums over sphere centers and over pairs of centers that the statistical properties of the bed of spheres must be considered.

The single summations in (2.6.4) are readily evaluated for a bed of randomly placed, freely overlapping solid spheres. The probability that an infinitesimal volume $d^3\boldsymbol{\rho}_1$ contains a sphere center is simply $n\,d^3\boldsymbol{\rho}_1$ where n is the average density of sphere centers, and

$$\left\langle \sum_i \nabla F_a(\boldsymbol{\rho}_i) \right\rangle = n \int_{\mathscr{V}} \nabla F_a(\boldsymbol{\rho}_1)\,d^3\boldsymbol{\rho}_1 \qquad (2.6.5)$$

The probability that a randomly selected point is in the void region is just the void fraction Φ, thus

$$\left\langle g(x) \sum_i \nabla F_a(\boldsymbol{\rho}_i) \right\rangle = \Phi n \int_{\mathscr{V}_a} \nabla F_a(\boldsymbol{\rho}_1)\,d^3\boldsymbol{\rho}_1 \qquad (2.6.6)$$

where $\mathscr{V}$ and $\mathscr{V}_a$ are very large volumes over which $\boldsymbol{\rho}_1$ is integrated. The presence of $g(x)$ in the integrand of (2.6.6) requires the probability that x be in the void, and that we find a sphere in $d^3\boldsymbol{\rho}_1$. A sphere of radius a located at the origin is excluded from the volume $\mathscr{V}_a$ in the integration of (2.6.6); since x must be a point in the void no sphere (of radius a) center can be at a distance less than a from the void point x.

Distinguishing between the two types of terms appearing in the double sum, namely those for which $i = j$ and those for which i and j are different, we have first an integral similar in form to (2.6.6)

$$\left\langle g(x) \sum_i [\nabla F_a(\boldsymbol{\rho}_i)]^2 \right\rangle = n\Phi \int_{\mathscr{V}_a} [\nabla F_a(\boldsymbol{\rho}_1)]^2\,d^3\boldsymbol{\rho}_1 \qquad (2.6.7)$$

If the centers are distributed at random regardless of the interpenetration of spheres, then the coordinates $\boldsymbol{\rho}_i$ and $\boldsymbol{\rho}_j$ of the members of the i, j-pair in the second integral from the double sum are completely uncorrelated and the double sum $i \neq j$ can be written

$$\left\langle g(x) \sum_{i \neq j} \nabla F_a(\boldsymbol{\rho}_i) \cdot \nabla F_a(\boldsymbol{\rho}_j) \right\rangle =$$

$$\Phi n^2 \int_{\mathscr{V}_a} \int_{\mathscr{V}_a} \nabla F_a(\boldsymbol{\rho}_1) \cdot \nabla F_a(\boldsymbol{\rho}_2)\,d^3\boldsymbol{\rho}_1\,d^3\boldsymbol{\rho}_2 \qquad (2.6.8)$$

The trial F_a taken from (2.6.3b) is substituted into the above integrals, (2.6.6) and (2.6.8) vanish, and the remaining integrals are:

$$\left\langle g \sum_i [\nabla F_a(\boldsymbol{\rho}_i)]^2 \right\rangle = n\Phi \int_{\mathscr{V}_a} \left[\frac{A}{\rho_1^3} - \frac{3(A \cdot \boldsymbol{\rho}_1)\boldsymbol{\rho}_1}{\rho_1^5} \right]^2 d^3\boldsymbol{\rho}_1$$

$$= \frac{8\pi n\Phi A^2}{3a^3} \qquad (2.6.9)$$

and

$$\left\langle \sum_i \nabla F_a(\boldsymbol{\rho}_i) \right\rangle = 4\pi nA/3 \qquad (2.6.10)$$

The upper bound (2.6.4) from the vanishing of (2.6.6) and (2.6.8) along with (2.6.9), and (2.6.10) can be written in terms of m

$$\frac{D_e}{D_0} < \Phi \left[\frac{1 + 8\pi nm^2/3a^3}{(1 + 4\pi nm/3)^2} \right].$$

$$(2.6.11)$$

The optimal value of m, obtained by differentiating the right side of (2.6.11) with respect to m, is

$$m = a^3/2.$$

$$(2.6.12)$$

When this value of m is substituted into (2.6.11) we have

$$\frac{D_e}{D_0} < \frac{\Phi}{\tau_1}$$

$$(2.6.13)$$

where

$$\tau_1 = 1 + 2\pi na^3/3$$

$$(2.6.14)$$

To eliminate n and a from the final result we use (2.6.2), and obtain

$$\tau_1 = 1 - \tfrac{1}{2} \ln \Phi.$$

$$(2.6.15)$$

The parameter τ_1 is related to the "tortuosity factor" which has been discussed by Carman (1956) and others. Although (2.6.15) was derived for a bed of uniform spheres, the same result is obtained when the bed contains overlapping spheres of several sizes. The generalized derivation is a straightforward repetition of the procedure used above for uniform spheres; this calculation is briefly discussed in Appendix 2.6.A. The result given by Eq. (2.6.15) may be compared with the earlier τ calculated for the isotropic suspension Eq. (2.4.17)

$$\tau_m = 1 + \tfrac{1}{2}(1 - \Phi)$$

$$(2.6.16)$$

We note, for example, that for dilute beds ($\Phi \to 1$) the two expressions are identical to first order in $(1 - \Phi)$. A graphical comparison of (2.6.15) and (2.6.16) is shown in Fig. 2.6.1. Both curves are theoretical lower bounds on $\tau (= \Phi D_0/D_e)$; the lower curve τ_m is the greatest lower bound that can be derived without considering statistical descriptions of the medium other than the void fraction, and the upper curve τ_1 illustrates, for a porous medium consisting of overlapping spheres, the improvement obtained by incorporating in the calculation a statistical description of the medium.

Calculated values of τ_1 are also compared in Fig. 2.6.1 with measured values of τ for packed beds of nonoverlapping solid spheres. The tortuosity parameters τ were obtained for various kinds of porous media among which were beds of uniform spheres and also mixtures of two or three different sphere sizes, the τ values were taken from Carman (1956).

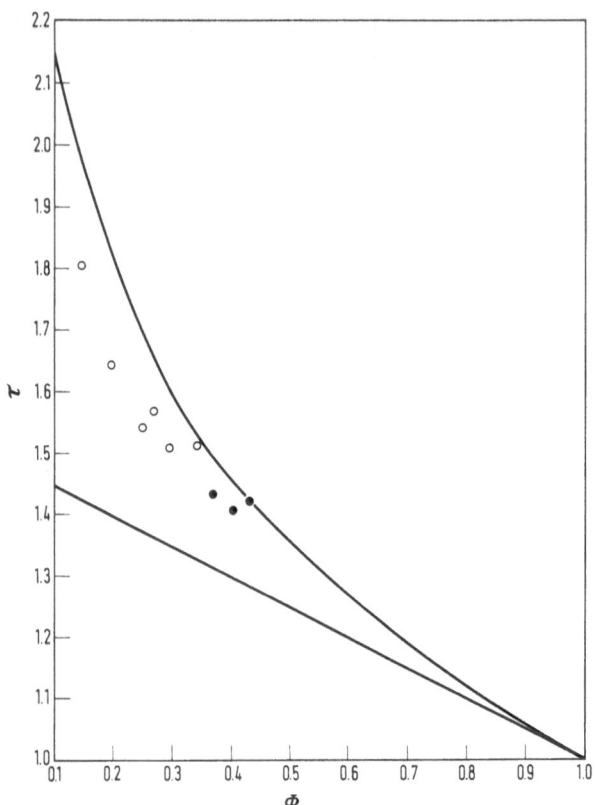

Fig. 2.6.1. Tortuosity parameter, $\tau = \Phi D_0/D_e$. Upper curve—(2.6.15), lower curve—(2.6.16). The plotted points represent experimental results for nonoverlapping spheres from Carman (1956): o-bed of uniform spheres, •-mixtures of spheres of more than one size. Both curves are theoretical lower bounds, the upper curve illustrates for a porous medium consisting of overlapping spheres, the improvement obtained by incorporating in the calculation a statistical description of the medium.

While this comparison of experimental data for nonoverlapping spheres with the theoretical estimate for overlapping spheres (upper curve) cannot be regarded as a valid test the latter, it does provide a good indication that the kind of rigorous bounds calculated here can furnish useful estimates of effective diffusion coefficients.

Appendix 2.6.A

Although (2.6.15) was derived for a bed of uniform spheres, it is interesting to note that the same result is obtained when the bed contains overlapping spheres of several sizes. The generalized derivation is a straightforward repetition of the procedure used for uniform spheres; we need only mention here the expression for the void fraction and the form of the trial function.

If our sphere bed is a superposition of separate randomly constructed beds for each sphere radius $(a_\alpha, a_\beta, \ldots)$, then the statistics for each of the radii are independent. For any one of the radii the sphere bed probability that a randomly chosen point is in the void is just the void fraction for that bed given by (2.6.2), and the obvious extension to a bed of over-lapping solid spheres of several sizes is

$$\Phi = \exp\left[-(4\pi/3)(n_\alpha a_\alpha^3 + n_\beta a_\beta^3 + \ldots)\right] \qquad (2.6.17)$$

where $n_\alpha, n_\beta, \ldots$ are the respective densities of sphere centers for the several sizes.

The trial function which replaces (2.6.3) is

$$\nabla u = A + m_\alpha \sum_i \nabla F_{a_\alpha}(\boldsymbol{\rho}_i) + m_\beta \sum_i \nabla F_{a_\beta}(\boldsymbol{\rho}_i) + \ldots \quad (2.6.18a)$$

where

$$F_{a_l}(\boldsymbol{\rho}_i) = \begin{cases} -A \cdot \boldsymbol{\rho}_i/\rho_i^3 & \rho_i > a_l \\ -A \cdot \boldsymbol{\rho}_i/a_l^3 & \rho_i < a_l \end{cases}$$

$$l = \alpha, \beta, \ldots \qquad (2.6.18b)$$

2.7. Knudsen Diffusion Through a Porous Medium

At sufficiently low pressures collisions between the molecules of the gas flowing through the porous material can be neglected and only collisions with the pore walls need be taken into account; this so-called Knudsen diffusion occurs when the mean free path for molecule-molecule collisions of the gas within the pores is large compared with the average pore diameter. (Refer to the discussion at the end of section 1.4.) One usually assumes that a molecule equilibrates with the wall at each collision, and that it "forgets" its previous history; thus molecules are re-emitted according to the laws of diffusive reflection.

Under these circumstances it is possible to formulate the flow of a gas through an arbitrary geometry in terms of an integral equation (De Marcus 1957, 1961). Such an equation can also be written for a random pore system, but our necessarily incomplete knowledge of the pore geometry in this case is reflected in a kernel which behaves sto-chastically. However, it is possible to apply the variational method of De Marcus (1957, 1961) to obtain a rigorous upper bound on the permeability (mean molecule flux per unit pressure gradient) of a porous medium in the Knudsen regime. This bound is expressed in terms of certain averages characterizing the random pore geometry and its derivation, together with an application to a model pore structure generated by randomly overlapping spheres, is the subject of the next three sections.

We consider only porous materials whose properties do not vary systematically with position. Thus average properties will not depend on position, and the material is statistically homogeneous. Of the many possible partial descriptions which might be developed for such a pore system, we shall find it most convenient to use four. These are: the void fraction Φ; the pore surface area s per unit volume; an average pore diameter $m_a = 4$ (void volume/pore wall surface area); and σ, the mean pore surface area which can be "seen" (i.e., reached by an unobstructed straight line) from a typical point on the void-solid interface. In addition, we define the probability $h_\sigma(\rho, n, n') \, d^3\rho \, d^2n \, d^2n'$ that two points x and x' on the pore wall surface that can see one another have unit normal vectors into the void n and n' falling in the elements of solid angle d^2n and d^2n', respectively, and a relative position vector $\rho = x' - x$ lying in the volume element $d^3\rho$. The probability functions were calculated in section 1.4 for a random bed of solid spheres.

To describe the molecular processes taking place within the porous medium, the pore wall surface is divided into small elements d^2x each located at some point x. The molecular diffusion depends on the number of molecules $\psi(x) \, d^2x$ re-emitted after pore wall collision from d^2x at x per unit time. It should be noted that ψ is defined only on the void-solid interface. To calculate ψ we follow De Marcus (1957, 1961) by introducing the probability $K(x, x') \, d^2x'$ that a molecule emitted from the pore wall at x will make its next wall collision within d^2x' at x'. The function $K(x, x')$ is also defined only when both x and x' lie on the void-solid interface, and, since we are assuming diffuse scattering at the walls, is given by the cosine law,

$$K(x, x') = K(x', x)$$
$$= \begin{cases} -\dfrac{[n(x) \cdot \rho][n(x') \cdot \rho]}{\pi \rho^4} & \text{(if } x \text{ can see } x') \\ 0 & \text{(otherwise)} \end{cases} \qquad (2.7.1)$$

where the unit normals $n(x)$ and $n(x')$ at points x and x' on the pore walls point into the void, and $\rho = (x' - x)$.

Now let us suppose the faces of a slab are the planes $x = 0$ and $x = L$. Let $\mathscr{S}$ be the void-solid interface inside the slab, and let $\mathscr{S}_0$ and $\mathscr{S}_L$ be the intersections of the planes $x = 0$ and $x = L$, respectively, with the void regions of the slab. Outside both ends of the slab we have a gas in equilibrium at the same temperatures, but a pressure difference

$P_L - P_0$ is maintained across the slab. The rates at which gas molecules enter the slab through unit areas of $\mathscr{S}_0$ and $\mathscr{S}_L$ are

$$\psi_0 = \frac{1}{4}\frac{\bar{v}}{k_B T}P_0 \quad \text{and} \quad \psi_L = \frac{1}{4}\frac{\bar{v}}{k_B T}P_L, \qquad (2.7.2)$$

$\bar{v}$ being the mean molecular speed. The definition of $K(x, x')$ can also be extended to include the case in which x or x' lies on $\mathscr{S}_0$ or $\mathscr{S}_L$ without necessitating any change in (2.7.1).

In the steady state, the condition that molecules must not accumulate on a unit element of area at any point x on $\mathscr{S}$ can be expressed in the form

$$\int_{\mathscr{S}_0 + \mathscr{S} + \mathscr{S}_L} K(x, x')\left[\psi(x') - \psi(x)\right]d^2x' = 0, \qquad (2.7.3)$$

together with the conditions

$$\begin{aligned}
\psi(x) &= \psi_0 \quad (x \text{ on } \mathscr{S}_0), \\
\psi(x) &= \psi_L \quad (x \text{ on } \mathscr{S}_L).
\end{aligned} \qquad (2.7.4)$$

Equation (2.7.3) is, in principle, sufficient to determine $\psi(x)$. In practice, of course, the complexity of the kernel $K(x, x')$ arising from its stochastic character prevents such an outright solution.

However, we are not interested in the detailed behavior of $\psi(x)$, but rather in the mean molecule flux J, which is the net rate at which gas molecules pass through unit total cross section of the slab, the impermeable solid portions being included in the averaging process. It is readily seen that J can be expressed as an integral of ψ

$$J = \frac{1}{V}\int_{\mathscr{S}_0 + \mathscr{S} + \mathscr{S}_L}\int_{\mathscr{S}_0 + \mathscr{S} + \mathscr{S}_L} K(x, x')\,\psi(x)\,(x' - x)\,d^2x\,d^2x' \qquad (2.7.5)$$

where V is the volume of the slab. The variational procedure of De Marcus now permits us to place an upper bound on the magnitude of J.

2.8. Variational Formulation for Knudsen Diffusion

Solution of the integral equation (2.7.3) subject to the condition (2.7.4) is entirely equivalent to minimizing the functional

$$\Gamma[\xi] = \frac{1}{2V}\int_{\mathscr{S}_0 + \mathscr{S} + \mathscr{S}_L}\int_{\mathscr{S}_0 + \mathscr{S} + \mathscr{S}_L} K(x, x')\{\xi(x') - \xi(x)\}^2\,d^2x\,d^2x' \qquad (2.8.1)$$

subject to the requirement that only trial functions $\xi(x)$ satisfying (2.7.4) are admissible. This is readily checked if a variation ξ_1 about the trial function ξ_0 is introduced into (2.8.1). Noting that any trial function satisfies (2.7.4) so that ξ_1 must vanish on $\mathscr{S}_0$ and $\mathscr{S}_L$, and also noting that, by (2.7.1), $K(x, x')$ is unchanged by interchange of x and x', we have for the term which is of the first order in the variation ξ_1,

$$\Gamma_1 = \frac{2}{V} \int_{\mathscr{S}} \xi_1(x) \int_{\mathscr{S}_0 + \mathscr{S} + \mathscr{S}_L} K(x, x') \{\xi_0(x) - \xi_0(x')\} \, d^2x \, d^2x'. \quad (2.8.2)$$

The expression Γ_1 is equal to zero if and only if the solution ψ of integral equation (2.7.3) is substituted into Eq. (2.8.2) in place of the trial function $\xi_0(x)$, and since Γ is positive for any $\xi(x)$ (or ξ_1)

$$\Gamma[\psi] \leq \Gamma[\xi] \quad (2.8.3)$$

We must now relate the mean molecular flux (2.7.5), to the minimum of Γ.

The simplest admissible trial function $\xi(x)$ is

$$\psi_m(x) = \psi_0 + \boldsymbol{\alpha} \cdot x \quad (2.8.4)$$

where

$$\boldsymbol{\alpha} = \frac{\psi_L - \psi_0}{L} i_x = \frac{1}{4} \frac{\bar{v}}{k_B T} \frac{P_L - P_0}{L} i_x$$

is the mean gradient in ψ, and i_x is a unit vector in the positive x-direction. All other trial functions can be written

$$\xi(x) = \psi_m(x) + \psi_f(x), \quad (2.8.5)$$

the trial fluctuation ψ_f being required to vanish on $\mathscr{S}_0$ and $\mathscr{S}_L$. In terms of $\psi_f(x)$ and the definitions

$$\Gamma_{mm} = \frac{1}{2V} \boldsymbol{\alpha}\boldsymbol{\alpha} : \int_{\mathscr{S}_0 + \mathscr{S} + \mathscr{S}_L} \int_{\mathscr{S}_0 + \mathscr{S} + \mathscr{S}_L} K(x, x')(x' - x)(x' - x) \, d^2x \, d^2x', \quad (2.8.6)$$

$$\Gamma_{mf}[\psi_f] = \frac{1}{2V} \boldsymbol{\alpha} \cdot \int_{\mathscr{S}_0 + \mathscr{S} + \mathscr{S}_L} \int_{\mathscr{S}_0 + \mathscr{S} + \mathscr{S}_L} \\ \cdot K(x, x')(x' - x) \{\psi_f(x') - \psi_f(x)\} \, d^2x \, d^2x', \quad (2.8.7)$$

$$\Gamma_{ff}[\psi_f] = \frac{1}{2V} \int_{\mathscr{S}_0 + \mathscr{S} + \mathscr{S}_L} \int_{\mathscr{S}_0 + \mathscr{S} + \mathscr{S}_L} \\ \cdot K(x, x') \{\psi_f(x') - \psi_f(x)\}^2 \, d^2x \, d^2x', \quad (2.8.8)$$

the functional (2.8.1) becomes

$$\Gamma[\xi] = \Gamma_{mm} + 2\Gamma_{mf}[\psi_f] + \Gamma_{ff}[\psi_f] \quad (2.8.9)$$

Now, if ψ_f is an admissible trial function and λ is some arbitrary constant, $\lambda\psi_f$ is also admissible. For a given choice of ψ_f we may minimize with respect to λ; the optimum value of λ so obtained is

$$\lambda_{\mathrm{opt}} = -\Gamma_{mf}[\psi_f]/\Gamma_{ff}[\psi_f] \qquad (2.8.10)$$

and the corresponding Γ is

$$\Gamma[\psi_m + \lambda_{\mathrm{opt}}\psi_f] = \Gamma_{mm} - \Gamma_{mf}^2[\psi_f]/\Gamma_{ff}[\psi_f] \qquad (2.8.11)$$

The least possible value of Γ is reached when ξ equals ψ, the solution of (2.7.3) and (2.7.4). If $\psi_f \equiv \psi - \psi_m$ is the associated fluctuation, minimization with respect to λ must give $\lambda_{\mathrm{opt}} = 1$. Thus, by (2.8.10),

$$\Gamma_{ff}[\psi - \psi_m] = -\Gamma_{mf}[\psi - \psi_m] \qquad (2.8.12)$$

Insertion of (2.8.12) into (2.8.9) and comparison with (2.7.5) yields

$$\Gamma[\psi] = \Gamma_{mm} + \Gamma_{mf}[\psi - \psi_m] = -\boldsymbol{J}\cdot\boldsymbol{\alpha} \qquad (2.8.13)$$

the last step being accomplished by invoking the symmetry of $K(x, x')$ with respect to the interchange of x and x'.

If ψ_f does not equal $\psi - \psi_m$ the resulting Γ will necessarily be larger than $\Gamma[\psi]$ or in view of Eq. (2.8.13)

$$\Gamma_{mm} - \Gamma_{mf}^2[\psi_f]/\Gamma_{ff}[\psi_f] > -\boldsymbol{J}\cdot\boldsymbol{\alpha} \qquad (2.8.14)$$

For every choice of ψ_f which vanishes on $\mathscr{S}_0$ and $\mathscr{S}_L$, the inequality (2.8.14) gives us a rigorous upper bound on the quantity $(-\boldsymbol{J}\cdot\boldsymbol{\alpha})$. For an isotropic material, or for an anisotropic material in which $\boldsymbol{\alpha}$ lies along one of the principal axes, the vectors $\boldsymbol{J}$ and $\boldsymbol{\alpha}$ will be parallel and have opposing directions; under these conditions the quantity bounded by (2.8.14) is just $J\alpha$, the product of the magnitudes.

2.9. Upper Bounds on the Knudsen Permeability

We now consider the simplest admissible trial function $\psi_f = 0$, Γ_{mf} and Γ_{ff} of expression (2.8.9) vanish, and the upper bound becomes simple Γ_{mm}. The evaluation of the integral (2.8.16) for Γ_{mm} is considerably simplified if we pass to the limit of a very thick slab by letting L become large compared to a typical pore dimension. In this limit the contributions from the surfaces $\mathscr{S}_0$ and $\mathscr{S}_L$ go to zero as L^{-1}, reflecting the fact that most of the pore surface of the slab cannot be seen from the entrance or exit faces. Only the integrations over $\mathscr{S}$ need, therefore, be retained, and (2.8.14) becomes

$$- \boldsymbol{J} \cdot \boldsymbol{\alpha} < \frac{1}{2V} \boldsymbol{\alpha}\boldsymbol{\alpha} : \int_{\mathscr{S}} \int_{\mathscr{S}} K(\boldsymbol{x}, \boldsymbol{x}')(\boldsymbol{x}' - \boldsymbol{x})(\boldsymbol{x}' - \boldsymbol{x}) \, d^2 x \, d^2 x' \quad (2.9.1)$$

with K given by (2.7.1).

Because of the stochastic character of $K(\boldsymbol{x}, \boldsymbol{x}')$ the integral in (2.9.1) cannot be calculated by the usual methods, but it can at least be interpreted as an average. If $\boldsymbol{\rho} = (\boldsymbol{x}' - \boldsymbol{x})$ is the displacement of a molecule between successive wall collisions, then

$$\frac{1}{V} \int_{\mathscr{S}} \int_{\mathscr{S}} K(\boldsymbol{x}, \boldsymbol{x}')(\boldsymbol{x}' - \boldsymbol{x})(\boldsymbol{x}' - \boldsymbol{x}) \, d^2 x \, d^2 x' = s \langle \boldsymbol{\rho}\boldsymbol{\rho} \rangle_{\mathscr{S}}. \quad (2.9.2)$$

The angular brackets $\langle \dots \rangle_{\mathscr{S}}$ in (2.9.2) denote a mean over a large number of randomly selected surface elements within the slab and s is the pore wall area per unit volume. So long as the quantity being averaged involves only $\boldsymbol{\rho}$ and the normals $\boldsymbol{n}$ and $\boldsymbol{n}'$ at $\boldsymbol{x}$ and $\boldsymbol{x}'$, this type of average can be expressed in terms of the probability density $h_\sigma(\boldsymbol{\rho}, \boldsymbol{n}, \boldsymbol{n}')$ and the mean pore surface area σ which can be seen from a typical point on the void-solid interface, both defined in section 2.7

$$\langle q(\boldsymbol{\rho}, \boldsymbol{n}, \boldsymbol{n}') \rangle_{\mathscr{S}} = -\frac{\sigma}{\pi} \int \int \int q(\boldsymbol{\rho}, \boldsymbol{n}, \boldsymbol{n}') h_\sigma(\boldsymbol{\rho}, \boldsymbol{n}, \boldsymbol{n}') \frac{(\boldsymbol{n} \cdot \boldsymbol{\rho})(\boldsymbol{n}' \cdot \boldsymbol{\rho})}{\rho^4} d^2 n \, d^2 n' \, d^3 \rho$$
$$(2.9.3)$$

the integrations extending over all $\boldsymbol{\rho}$ (an infinite volume), and all orientations of the unit vectors $\boldsymbol{n}$ and $\boldsymbol{n}'$.

In contrast to the integral in (2.9.1) the integral in (2.9.3) does not involves only $\boldsymbol{\rho}$ and the normals $\boldsymbol{n}$ and $\boldsymbol{n}'$ at $\boldsymbol{x}$ and $\boldsymbol{x}'$, this type of average For the case of randomly overlapping solid spheres the form of h_σ has been calculated in section 1.4 and is given by (1.4.4). In view of the isotropy of the pore geometry, the upper bound becomes

$$\frac{J}{\alpha} < \frac{1}{6} s \langle \rho^2 \rangle_{\mathscr{S}} = \frac{16}{3} \frac{\Phi^2}{s} \quad (2.9.4)$$

The first form of inequality (2.9.4) is, in fact, true for any isotropic material; it is strongly reminiscent of the Einstein relation between a diffusion coefficient and the mean square displacement during a single step in a Brownian motion. Indeed if we were to regard a molecule moving through the pore system as a Brownian particle we would obtain J/α equal to $\frac{1}{6} s \langle \rho^2 \rangle_{\mathscr{S}}$. But every step of a Brownian motion is isotropic (that is, all directions are equally likely), whereas the local anisotropies in the motion of a molecule in Knudsen flow are determined by the orientation of the pore surface at the point of the last wall collision. This turns the equality of the Einstein relation into an inequality.

To improve on the bound (2.9.4) we must introduce a ψ_f which has some relation to the random pore geometry, i.e., ψ_f should show stochastic behavior. A very simple way of achieving this is to use

$$\psi_f(x) = \begin{cases} \boldsymbol{\alpha} \cdot \boldsymbol{n}(x) & (x \text{ on } \mathscr{S}) \\ 0 & (x \text{ on } \mathscr{S}_0 \text{ or } \mathscr{S}_L) \end{cases} \tag{2.9.5}$$

In the limit of a very thick slab, this leads, by the same reasoning as before, to the inequality

$$-\boldsymbol{J} \cdot \boldsymbol{\alpha} < \frac{1}{2}s\left[\boldsymbol{\alpha}\boldsymbol{\alpha} : \langle \boldsymbol{\rho}\boldsymbol{\rho} \rangle_{\mathscr{S}} - \frac{\{\boldsymbol{\alpha}\boldsymbol{\alpha} : \langle \boldsymbol{\rho}(\boldsymbol{n}' - \boldsymbol{n}) \rangle_{\mathscr{S}}\}^2}{\boldsymbol{\alpha}\boldsymbol{\alpha} : \langle (\boldsymbol{n}' - \boldsymbol{n})(\boldsymbol{n}' - \boldsymbol{n}) \rangle_{\mathscr{S}}} \right] \tag{2.9.6}$$

which becomes

$$\frac{J}{\alpha} < \frac{1}{6}s\left[\langle \rho^2 \rangle_{\mathscr{S}} - \frac{\langle \boldsymbol{\rho} \cdot (\boldsymbol{n}' - \boldsymbol{n}) \rangle_{\mathscr{S}}^2}{\langle (\boldsymbol{n}' - \boldsymbol{n}) \cdot (\boldsymbol{n}' - \boldsymbol{n}) \rangle_{\mathscr{S}}} \right] \tag{2.9.7}$$

for an isotropic material. In the case of a pore system generated by randomly overlapping spheres the averages appearing in (2.9.7) may be evaluated through (1.4.4) and (2.9.3), to give

$$\frac{J}{\alpha} < \frac{48}{13}\frac{\Phi^2}{s} \tag{2.9.8}$$

Although the upper bound (2.9.8) has the same form as (2.9.4), the numerical coefficient has been reduced by a factor of about 2/3.

In terms of the permeability $\Pi = JL/|P_L - P_0|$ (i.e., the mean molecule flux per unit pressure gradient), (2.9.8) can be rewritten

$$\Pi < \frac{12}{13}\frac{\bar{v}}{k_B T}\frac{\Phi^2}{s} \tag{2.9.9}$$

This result has been proposed by Derjaguin (1946) as an equality, but Lassettre (1958) has shown that the right-hand side of (2.9.9) only approaches the true permeability in the limit of a very dilute bed of spheres ($\Phi \to 1$). For more realistic values of Φ, Derjaguin's permeability estimate tends to run about 40% too high for a wide variety of porous media. (Carman, 1956).

Chapter 3

Diffusion Limited Reactions

3.1. Introduction

A diffusion limited reaction rate is one which depends solely on the rate at which solute molecules diffuse to the site of reaction such as the surface of a reactant particle. When all reaction processes at the surface of the reactant particle are so fast that surface equilibrium conditions are maintained, the effective (or net) reaction rate does not depend on the nature of the process at the site of reaction, but rather upon the rate of transport by diffusion.

The first application of a variational principle is to determine the rate at which a solute precipitates from a homogeneous solution. The precipitate particles are modeled as spheres which are all of the same radius and arranged in a simple cubic lattice in the solution. The solution of this problem is mathematically difficult because of the geometrical disparity of the boundaries of the sphere and the cubic cell, but a variational principle can be used to estimate the rate of precipitation.

The second application of a variational procedure is to establish rigorous upper and lower bounds on the rate of diffusion limited reaction, in which an excited species B^* is generated at a constant and uniform rate throughout a solution which also contains a quenching species A_q. The de-excitation or quenching is assumed to occur instantaneously whenever a B^* molecule has diffused within a certain critical distance a of a quencher particle. At higher quencher concentration the A_q particles do not act independently, for the diffusion of B^* to one A_q particle will be affected by the presence of other particles in its vicinity. Moreover the distribution of the A_q particles is only known in a probabilistic sense, and as in the case of the diffusion problem of chapter 2, an exact calculation of the surface reaction rate is extremely difficult. Variational upper and lower bounds on the effective reaction rate constant are calculated.

3.2. Diffusion Limited Precipitation

In a series of articles Ham (1958, 1959) used variational methods to calculate upper bounds on precipitation rates from solution. In the example which we shall study, Ham assumed that each spherical precipitate particle (of radius a) would be found at the center of a unit cube of an infinite cubic lattice structure with no vacant sites (Fig. 3.2.1).

The calculations are simplified by the physical fact that the slowest normal mode dominates the decay after a short initial period. Further, the flux vanishes on the outer walls of the unit cube providing a natural boundary condition there, hence the trial functions for the variational principle need not satisfy the boundary conditions on the walls of the cube. As a consequence, one can replace the outer boundary condition by a boundary condition on an outer sphere of the same volume concentric with the center of the precipitate sphere. By solving the diffusion equation for this much simpler problem we obtain trial functions.

Consider the diffusion limited precipitation from a supersaturated solution. The solute concentration $c(x, t)$ as a function of position x and time t is given at every point in the solution by the diffusion equation

$$\frac{\partial}{\partial t} c(x, t) = D \nabla^2 c(x, t) \tag{3.2.1}$$

with constant diffusion coefficient D. The reaction rate at the surface of the precipitate particle is assumed to be infinitely fast, and the concentration of solute at the surface $\mathscr{S}$ of a precipitate particle is the equilibrium or saturated value c_s,

$$c(x, t) = c_s \quad \text{on } \mathscr{S}. \tag{3.2.2}$$

We now divide the solution into equivalent cubic cells and place a spherical precipitate particle of radius a at the center of each cell in the cubic lattice as in Fig. 3.2.1. If the initial concentration is uniform, say

$$c(x, 0) = c_0 + c_s, \tag{3.2.3}$$

then the concentration profile will preserve the lattice symmetry during the precipitation. The normal component of the flux $\mathbf{j}(x, t)$ will vanish at the surface $\mathscr{S}_c$ of any cell

$$n \cdot \mathbf{j}(x, t) = -Dn \cdot \nabla c(x, t) = 0 \quad \text{on } \mathscr{S}_c, \tag{3.2.4}$$

where n is the outward normal to the cubic cell wall. If the density of the solute in the precipitate is much larger than the solute concentration in solution in excess of the saturated value, then the radius of the precipitate particle changes very slowly and we will assume a is nearly

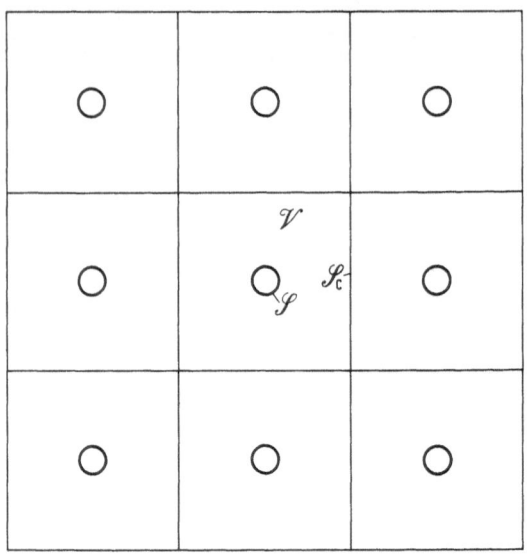

Fig. 3.2.1. A simple periodic lattice of spherical precipitate particles.

constant in time. Moreover, the dimensions of a particle are to be considered small in comparison to the distance between particles

$$a/r_s \ll 1, \tag{3.2.5}$$

where r_s is the radius of a sphere of the same volume as the cubical cell. Thus

$$\tfrac{4}{3}\pi r_s^3 n = 1, \tag{3.2.6}$$

n being the number of precipitate particles per unit volume.

It suffices to solve (3.2.1) for a single cell subject to the boundary conditions (3.2.2), (3.2.4), and the initial value (3.2.3). To proceed we expand $c(x, t) - c_s$ in eigenfunctions of the equation

$$\nabla^2 \psi_p(x) + \lambda_p^2 \psi_p(x) = 0 \quad \text{in } \mathscr{V} \tag{3.2.7}$$

and

$$\psi_p(x) = 0 \quad \text{on } \mathscr{S}, \tag{3.2.8}$$

$$n \cdot \nabla \psi_p(x) = 0 \quad \text{on } \mathscr{S}_c \tag{3.2.9}$$

that is

$$c(x, t) - c_s = \sum_{p=0}^{\infty} a_p(t) \psi_p(x). \tag{3.2.10}$$

Substituting (3.2.10) into (3.2.1) and using (3.2.7) we find

$$\frac{da_p(t)}{dt} = -D\lambda_p^2 a_p(t) \tag{3.2.11}$$

as a result of the orthonormality of the $\psi_p(x)$,

$$\int_{\mathscr{V}} \psi_p(x)\,\psi_{p'}(x)\,d^3x = \delta_{pp'}, \tag{3.2.12}$$

where the integral is taken over the interior of the cell excluding the volume of the particle. Hence

$$c(x, t) = c_s + \sum_{p=0}^{\infty} a_p(0)\,e^{-t/\tau_p}\psi_p(x) \tag{3.2.13}$$

where

$$\tau_p = (\lambda_p^2 D)^{-1}, \tag{3.2.14}$$

and

$$a_p(t) = \int_{\mathscr{V}} \{c(x, t) - c_s\}\,\psi_p(x)\,d^3x \tag{3.2.15}$$

The solution of this problem for $\psi_p(x)$ is mathematically difficult because of the geometrical disparity of the boundaries of the interior spherical particle and the outer boundaries of the cubic cell.

3.3. The Spherical Cell Approximation

Evaluation of ψ_p and λ_p is greatly simplified if we replace the cubic cell by a sphere of equivalent volume, whose radius r_s is given by (3.2.6). For spherically symmetric solutions of the eigenfunction equation

$$\psi_p = \left(\frac{A_p}{\lambda_p}\right) \frac{\sin \lambda_p(r_\sigma - a)}{r_\sigma} \ , \tag{3.3.1}$$

A_p is a normalizing constant,

$$\left(\frac{A_p}{\lambda_p}\right)^2 = \frac{1}{4\pi} \left\{ \frac{2(1 + \lambda_p^2 r_s^2)}{\lambda_p^2 r_s^2 (r_s - a) - a} \right\} , \tag{3.3.2}$$

and r_σ is the radial distance from the center of the precipitate particle. From (3.2.15) for the initial distribution $c_0 + c_s$ we find

$$a_p(0) = 4\pi c_0 a A_p / \lambda_p^2 \tag{3.3.3}$$

The eigenvalues λ_p are the roots of the equation

$$\tan\left[\lambda_p(r_s - a)\right] = \lambda_p r_s \tag{3.3.4}$$

obtained by applying boundary condition (3.2.4) to the spherical outer cell boundary. Since a/r_s is small we can expand the tangent and show that the smallest non trivial root of (3.3.4) is

$$r_s^2 \lambda_0^2 = \frac{3a}{r_s}\left[1 + 0\left(\frac{a}{r_s}\right)\right] \qquad (3.3.5)$$

The second root is given approximately by $r_s \lambda_1 = 4.5$, and all the other eigenvalues are greater than λ_1. Thus for small a/r_s,

$$\tau_0 \gg \tau_p \quad \text{for all } p \geq 1. \qquad (3.3.6)$$

Moreover (3.3.2) and (3.3.3) show that for the coefficients in (3.2.13)

$$a_0 \gg a_p \quad \text{for all} \quad p \geq 1. \qquad (3.3.7)$$

Thus when $t > \tau_1$ only the zeroth normal mode continues to make a significant contribution to the precipitation.

The eigenfunction ψ_0 given by (3.3.1) is not strongly dependent on r_σ; in fact

$$\psi_0 = A_0\left(1 - \frac{a}{r_\sigma} - \frac{ar_\sigma^2}{2r_s^3} + \cdots \right) \qquad (3.3.8)$$

We therefore have for $r_\sigma \gg a$ and $t \gg \tau_1$

$$c(\mathbf{x}, t) - c_s = c_v(t) = c_0 e^{-t/\tau_0} \qquad (3.3.9)$$

whereas for small r_σ and $t \gg \tau_1$

$$c(\mathbf{x}, t) - c_s = c_v(t)\left(1 - \frac{a}{r_\sigma} + \cdots \right) \qquad (3.3.10)$$

We end this section with a final note on the "isolated particle approximation." At infinite dilution of precipitate particle the quantity a/r_s is in fact extremely small, each precipitate sphere is in effect surrounded by an infinite volume, and they act independently of one another. Then (3.3.9) and (3.3.10) provide the true concentration distribution about each precipitate sphere, and the reaction rate at infinite dilution, τ_0^{-1}, can be obtained directly from (3.3.10)

$$\tau_0^{-1} = 4\pi a D n = 3 D a r_s^{-3}, \qquad (3.3.11)$$

by calculating the solute flux at the precipitate surface.

3.4. Variational Principle and Upper Bound on the Precipitation Rate

The constant τ_0^{-1} is a measure of the rate of precipitation. By (3.2.14) τ_0 is given in terms of the lowest eigenvalue, which can be approximated by variational methods.

Consider the variational integral

$$K(u) = \left[\int_{\mathcal{V}} \nabla u \cdot \nabla u \, d^3 x \right] \Big/ \left[\int_{\mathcal{V}} u^2 \, d^3 x \right] \tag{3.4.1}$$

where the integrals are over the cubic volume $\mathcal{V}$ of the lattice excluding the spherical precipitate particle. The trial function $u(x)$ must satisfy (3.2.2) at the precipitate surface, but need not satisfy the boundary condition (3.2.4) on the outer wall of the cubic cell.

Clearly as the integrals in (3.4.1) are positive $K(u)$ must take on a non zero minimum. Furthermore the function u_0 which minimizes K, also causes the first variation to vanish. If $u = u_0 + u_1$, the term linear in u_1 is

$$K_1 = \left[2 \int_{\mathcal{V}} \{ \nabla u_1 \cdot \nabla u_0 - \gamma_0 u_1 u_0 \} \, d^3 x \right] \left[\int_{\mathcal{V}} u_0^2 \, d^3 x \right]^{-1} \tag{3.4.2}$$

where

$$\gamma_0 = K_0 = K(u_0) \tag{3.4.3}$$

The divergence theorem applied to the first term in (3.4.2) gives

$$K_1 = 2 \left[\int_{\mathcal{S} + \mathcal{S}_c} u_1 \nabla u_0 \cdot n \, d^2 x - \int_{\mathcal{V}} u_1 \{ \nabla^2 u_0 + \gamma_0 u_0 \} \, d^3 x \right] \\ \cdot \left[\int_{\mathcal{V}} u_0^2 \, d^3 x \right]^{-1} \tag{3.4.4}$$

where n is the outward normal to the cell volume $\mathcal{V}$.
If we note that

$$u = u_0 = 0 \quad \text{on } \mathcal{S} \tag{3.4.5}$$

and require that

$$\nabla u_0 \cdot n = 0 \quad \text{on } \mathcal{S}_c \tag{3.4.6}$$

then K_1 vanishes if and only if

$$\nabla^2 u_0 + \gamma_0 u_0 = 0. \tag{3.4.7}$$

Then u_0 is a solution of the eigenvalue problem (3.2.7) subject to boundary conditions (3.2.8) and (3.2.9). Further, the minimum of $K(u)$, γ_0 must be the lowest eigenvalue λ_0^2 of (3.2.7) and

$$\lambda_0^2 = K_0 \le K(u) \tag{3.4.8}$$

The error in λ_0^2 will depend on how closely u approximates ψ_0.

We shall use our previous discussion of the spherical cell to select the trial function. For a lowest order term in λ_0^2, it is sufficient to approximate $u(x)$ by a function satisfying the boundary condition (3.2.8) and Laplace's equation

$$\nabla^2 u(x) = 0. \tag{3.4.9}$$

Then, applying the divergence theorem to (3.4.1) we have

$$K(u) = \left[\int_{\mathscr{S}_c} u(x) \left\{ n \cdot \nabla u(x) \right\} d^2x \right] \bigg/ \left[\int_{\mathscr{V}} u^2(x) d^3x \right] \tag{3.4.10}$$

If $u(x) = c_v$ is nearly constant except in the immediate neighborhood of the precipitate particle, as (3.3.9) and (3.3.10) imply for the spherical cell approximation, the denominator in (3.4.10) is equal to Vc_v^2. Thus to an approximation which would neglect terms of order a/r_s,

$$K(u) = \left[\int_{\mathscr{S}_c} n \cdot \nabla u(x) d^2x \right] \bigg/ Vc_v. \tag{3.4.11}$$

From (3.4.9) the flux is a constant across any closed surface enclosing the precipitate particle, and the appropriate trial function is

$$u(x)/c_v = 1 - a/r_\sigma, \tag{3.4.12}$$

This leads to the approximation

$$K(u) = (3a/r_s^3) \left[1 + 0(a/r_s) \right] = \lambda_0^2 \tag{3.4.13}$$

This result agrees with (3.3.5), and when written in terms of $n \left[= 3/(4\pi r_s^3) \right]$ from (3.2.6) and $\tau_0^{-1} \left[= \lambda_0^2 D \right]$ from (3.2.14) we find that the lowest order term in (3.4.13) is equivalent to the "isolated particle approximation" (3.3.11). The second order term in a/r_s has been calculated

$$K(u) = \frac{3a}{r_s^3} \left[1 + R_a \frac{a}{r_s} + 0\left(\frac{a^2}{r_s^2} \right) \right] \tag{3.4.14}$$

where

$$R_a = -\left(\frac{3}{4\pi} \right)^{1/3} \left\{ \frac{3\sqrt{2}}{\pi} \arctan\left(\frac{\sqrt{2}}{2} \right) + 4 \left[\pi - 6 \ln(2 + \sqrt{3}) \right] \right\}$$

$$= 11.3 \ldots \tag{3.4.15}$$

The first term in (3.4.14) still gives the isolated particle value of λ_0^2 for infinite dilution of precipitate particles. The higher terms of the expansion in powers of the cube root of the density represent corrections due to the interaction of the concentration gradients around each precipitate particle, i.e., the effects that a solute molecule diffusing toward a spherical precipitate particle will experience due to the presence of neighboring precipitate particles.

3.5. Diffusion Controlled Quenching

A variational procedure was developed by R. Reck, G. Reck, and S. Prager to establish rigorous upper and lower bounds on the rate of diffusion controlled reaction, in which an excited species B^* is generated at a constant and uniform rate σ^* throughout a solution containing a quenching species A_q. De-excitation is assumed to occur instantaneously whenever a B^* molecule has diffused within a distance a of a quencher particle. The quenching ability of A_q is unaffected by its previous history and the spontaneous de-excitation of B^* is neglected. The quencher particles are assumed to be stationary.

At higher quencher concentrations, it is not possible to think of the A_q particles as acting independently, in that their reaction zones will overlap, and the diffusion of B^* to one A_q particle will be affected by the presence of other A_q particles in the vicinity. Bounds are obtained in terms of various statistical characterizations of the distribution of A_q particles in the solution, and explicit calculations are carried out for quencher particles which exert no forces on one another.

Let $\mathscr{V} - \hat{\mathscr{V}}$ be the reaction zone presented jointly by all the A_q particles in a representative volume $\mathscr{V}$ of solution, and let $\hat{\mathscr{V}}$ represent all points in $\mathscr{V}$ not within a distance a of some quencher particle. Let $\partial\hat{\mathscr{V}}$ represent the interface between $\mathscr{V} - \hat{\mathscr{V}}$ and $\hat{\mathscr{V}}$. (Fig. 3.5.1).

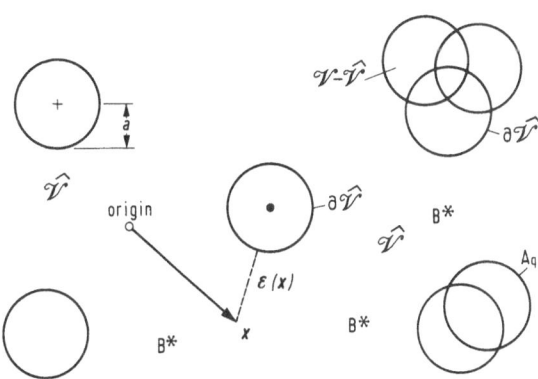

Fig. 3.5.1. Representation of interpenetrating sphere model of reaction zones. B^* is the excited species; A_q is the quencher species; $\hat{\mathscr{V}}$ is the void region; $\mathscr{V} - \hat{\mathscr{V}}$ is the reaction zone; x is the position vector of a point outside the reaction zone; $\partial\hat{\mathscr{V}}$ is the reactive interface between $\hat{\mathscr{V}}$ and $\mathscr{V} - \hat{\mathscr{V}}$; and $\varepsilon(x)$ is the minimum distance between x and $\partial\hat{\mathscr{V}}$.

The steady state concentration $c(x)$ of B^* at a point x in $\mathscr{V}$ must satisfy the diffusion equation when x is in $\mathscr{\hat{V}}$

$$D \nabla^2 c + \sigma^* = 0. \tag{3.5.1}$$

$c(x)$ will vanish if x is in $\mathscr{V} - \mathscr{\hat{V}}$ and the boundary condition for (3.5.1) is

$$c = 0 \quad \text{on } \partial\mathscr{\hat{V}}. \tag{3.5.2}$$

In the steady state a given rate of generation σ^* of B^* will maintain a mean concentration level $\langle c \rangle$ of excited species,

$$V \langle c \rangle = \int_{\mathscr{\hat{V}}} c(x)\, d^3x, \tag{3.5.3}$$

such that the overall rates of formation and destruction exactly compensate one another. For a particular quencher concentration $[A_q]$, σ^* will be proportional to $\langle c \rangle$, the proportionality coefficient k_e is an effective first order rate constant,

$$\sigma^* = k_e([A_q]) \langle c \rangle \tag{3.5.4}$$

3.6. Upper Bound on k_e

To obtain an upper bound on k_e we will show that the selection of the trial function u which minimizes the integral

$$\Upsilon(u) = \frac{D}{V} \int_{\mathscr{\hat{V}}} (\nabla u)^2 \, d^3x, \tag{3.6.1}$$

subject to the subsidiary condition,

$$V \langle c \rangle = \int_{\mathscr{\hat{V}}} u(x) \, d^3x \tag{3.6.2}$$

and the boundary condition

$$u(x) = 0 \quad \text{on } \partial\mathscr{\hat{V}} \tag{3.6.3}$$

is equivalent to the solution of (3.5.1) and (3.5.2). The subsidiary condition is introduced into the variational integral with a Lagrangian multiplier ω, and in place of Υ the functional

$$\Upsilon'(u) \equiv V^{-1} \int_{\mathscr{\hat{V}}} [D(\nabla u)^2 + \omega u(x)] \, d^3x - \omega \langle c \rangle \tag{3.6.4}$$

will be minimized subject only to (3.6.3). The $u(x)$ which gives this minimum, say u_0, must cause the variation of Υ' to vanish for any variation $u_1(x)$ which is zero on $\partial\mathscr{\hat{V}}$. But the terms linear in u_1 are

$$\Upsilon_1' = V^{-1} \int_{\hat{\mathscr{V}}} \left[2D\, \nabla u_0 \cdot \nabla u_1 + \omega u_1 \right] d^3 x \tag{3.6.5}$$

$$= V^{-1} \int_{\hat{\mathscr{V}}} \left[-2D\, \nabla^2 u_0 + \omega \right] u_1 \, d^3 x = 0. \tag{3.6.6}$$

Thus if Υ' is to be minimized u_0 must satisfy the Euler-Lagrange equation

$$-2D\, \nabla^2 u_0 + \omega = 0. \tag{3.6.7}$$

Equation (3.6.7) is equivalent to (3.5.1) if

$$\omega = -2\sigma^* \tag{3.6.8}$$

Since u must satisfy boundary condition (3.6.3) the minimizing $u(=c)$ is the solution of Eqs. (3.5.1) and (3.5.2).

The minimum value of Υ is closely related to the effective rate constant k_e. If $c(x)$ is a solution of (3.5.1) and (3.5.2) and λ is an adjustable parameter then

$$u(x) = \lambda c(x) \tag{3.6.9}$$

is a legitimate trial function and the optimal value of λ must be unity. Substitution of (3.6.9) into (3.6.4) together with the requirement that the resulting $\Upsilon'(\lambda)$ must have a minimum at $\lambda = 1$, leads to

$$\Upsilon_{\min} \equiv \frac{D}{V} \int_{\hat{\mathscr{V}}} (\nabla c)^2 \, d^3 x = V^{-1} \int_{\hat{\mathscr{V}}} \sigma^* c \, d^3 x = \sigma^* \langle c \rangle \tag{3.6.10}$$

A trial concentration u differing from c, when substituted into (3.6.1) will lead to a value of Υ that is greater than $\sigma^* \langle c \rangle$; since $\langle c \rangle$ is fixed by the subsidiary condition (3.6.2) this is an upper bound on k_e,

$$k_e = \Upsilon_{\min}/\langle c \rangle^2 \le D \left[V^{-1} \int_{\hat{\mathscr{V}}} (\nabla u)^2 \, d^3 x \right] \bigg/ \left[V^{-1} \int_{\hat{\mathscr{V}}} u \, d^3 x \right]^2 \tag{3.6.11}$$

While the upper bound in this section could be obtained by a method similar to that of section 3.4, a more elaborate derivation is used here to display the relationship between the upper and lower bound reciprocal variational principles.

A class of trial functions suitable for substituting into (3.6.11) may be constructed through use of the parameter $\varepsilon(x)$, the minimum distance to $\partial \hat{\mathscr{V}}$ from a point x in $\hat{\mathscr{V}}$ outside the reaction zone (Fig. 3.5.1),

$$u(x) = H^*(\varepsilon) \tag{3.6.12}$$

Equation (3.6.3) will be satisfied by any $H^*(\varepsilon)$ which vanishes at $\varepsilon = 0$. Note that H^* is a perfectly ordinary non-stochastic function of ε, however, $u(x)$ is stochastic due to the nature of the dependence of ε on x.

If $\varepsilon(x)$ is the minimum distance from x to the interface $\partial\hat{\mathscr{V}}$, then this distance must lie on a path in the radial direction from the nearest sphere. The gradient of any function is a vector whose magnitude is equal to the maximum change of that function and which points in the direction of that change. Then we have

$$\nabla\varepsilon(x) = n \qquad (3.6.13)$$

where n is a unit vector directed radially from the nearest sphere. Using this result we can carry out the substitution of H^* into (3.6.11) in terms of the probability $P(\varepsilon)\,d\varepsilon$ that a randomly selected point in $\mathscr{V}$ is in $\hat{\mathscr{V}}$ and at a distance between ε and $\varepsilon + d\varepsilon$ from the closest point on $\partial\hat{\mathscr{V}}$

$$k_e < D \int_0^\infty P(\varepsilon)\left(\frac{dH^*}{d\varepsilon}\right)^2 d\varepsilon \left/ \left[\int_0^\infty P(\varepsilon)\,H^*(\varepsilon)\,d\varepsilon\right]^2\right. . \qquad (3.6.14)$$

The function $H^*(\varepsilon)$ which gives the best upper bound must satisfy the Euler-Lagrange differential equation

$$\frac{d}{d\varepsilon}\left[P(\varepsilon)\frac{dH^*}{d\varepsilon}\right] + P(\varepsilon) = 0 \qquad (3.6.15)$$

subject to the conditions

$$H^*(0) = 0 \qquad (3.6.16)$$

$$P\,dH^*/d\varepsilon \to 0 \quad\text{as}\quad \varepsilon \to \infty \qquad (3.6.17)$$

This may be solved by quadratures to give

$$H^*(\varepsilon) = \int_0^\varepsilon \int_{\varepsilon'}^\infty [P(\varepsilon')]^{-1}\,P(\varepsilon'')\,d\varepsilon'\,d\varepsilon''. \qquad (3.6.18)$$

Upon substitution of this result into (3.6.14) we have

$$k_e < D \left/ \int_0^\infty \int_0^\varepsilon \int_{\varepsilon'}^\infty P(\varepsilon)\,P(\varepsilon'')\,[P(\varepsilon')]^{-1}\,d\varepsilon\,d\varepsilon'\,d\varepsilon''\right. \qquad (3.6.19)$$

3.7. Lower Bound on k_e

For the lower bound on the effective rate constant we reformulate the variational problem in terms of a trial flux vector, $q(x)$, and require the minimization of

$$\Xi = V^{-1} \int_{\hat{\mathscr{V}}} D^{-1}[q(x)]^2\,d^3x \qquad (3.7.1)$$

subject to the subsidiary condition

$$\nabla \cdot \boldsymbol{q} = \sigma^* \qquad (\boldsymbol{x} \text{ in } \hat{\mathscr{V}}) \tag{3.7.2}$$

The subsidiary condition is introduced by means of a Lagrangian multiplier $\beta(\boldsymbol{x})$, and we consider

$$\theta = \Xi' \equiv V^{-1} \int_{\hat{\mathscr{V}}} \{D^{-1}[\boldsymbol{q}(\boldsymbol{x})]^2 + \beta(\boldsymbol{x})(\nabla \cdot \boldsymbol{q}(\boldsymbol{x}) - \sigma^*)\} d^3x \tag{3.7.3}$$

The trial flux $\boldsymbol{q}_0$ that minimizes Ξ' must make the first variation of Ξ' vanish. But if $\boldsymbol{q} = \boldsymbol{q}_0 + \boldsymbol{q}_1$ the terms linear in $\boldsymbol{q}_1$ are

$$\Xi'_1 = V^{-1} \int_{\hat{\mathscr{V}}} \{2D^{-1}\boldsymbol{q}_0 \cdot \boldsymbol{q}_1 + \beta(\boldsymbol{x}) \nabla \cdot \boldsymbol{q}_1\} d^3x = 0 \tag{3.7.4}$$

The second term in the integrand is transformed by Gauss' theorem to give

$$\Xi'_1 = V^{-1} \int_{\hat{\mathscr{V}}} \{2D^{-1}\boldsymbol{q}_0 - \nabla \beta(\boldsymbol{x})\} \cdot \boldsymbol{q}_1 \, d^3x$$
$$+ V^{-1} \int_{\partial \hat{\mathscr{V}}} \beta(\boldsymbol{x}) \boldsymbol{q}_1 \cdot \boldsymbol{n} \, d^2x \tag{3.7.5}$$

where $\boldsymbol{n}$ is the outward normal to $\hat{\mathscr{V}}$.

The variation $\boldsymbol{q}_1$ is not restricted anywhere, not even on the surface $\partial \hat{\mathscr{V}}$, and the Euler-Lagrange equations for the minimization of Ξ subject to the condition (3.7.2) are therefore

$$\boldsymbol{q}_0(\boldsymbol{x}) = \tfrac{1}{2} D \nabla \beta(\boldsymbol{x}) \quad \text{in } \hat{\mathscr{V}} \tag{3.7.6}$$

$$\beta(\boldsymbol{x}) = 0 \qquad \text{on } \partial \hat{\mathscr{V}} \tag{3.7.7}$$

Equations (3.7.6) and (3.7.7) together with (3.7.2) are equivalent to the diffusion equation (3.5.1) and boundary condition (3.5.2), if we make the identification

$$\beta(\boldsymbol{x}) = -2c(\boldsymbol{x}) \tag{3.7.8}$$

The true flux and concentration distributions are related by the equation

$$\mathbf{j}(\boldsymbol{x}) = -D \nabla c(\boldsymbol{x}) \tag{3.7.9}$$

and the minimum value of Ξ must therefore equal the minimum value of Υ, namely $\sigma^* \langle c \rangle$. The difference in the variational principles is that Υ is minimized with constant $\langle c \rangle$, whereas Ξ is minimized with constant σ^*. In other words, while the variational principle (3.6.1) of the preceding discussion places an upper bound on $\sigma^* \langle c \rangle$ for a given value of $\langle c \rangle$, the present principle places an upper bound on $\sigma^* \langle c \rangle$ for a given value of σ^*. The former, as we have seen, is equivalent to an upper bound on k_e, but the latter corresponds to an upper bound on $1/k_e$ or to a *lower* bound on k_e,

$$k_e = \frac{\sigma^{*2}}{\Xi_{\min}} \geq \frac{\sigma^{*2}}{\Xi} = \sigma^{*2} \Big/ \left[\frac{1}{V} \int_{\hat{\mathscr{V}}} D^{-1} q^2(x) \, d^3x \right] . \quad (3.7.10)$$

A simple trial function which will satisfy the subsidiary condition (3.7.2) is

$$q(x) = \frac{1}{4\pi} \int_{\mathscr{V}} h^*(x') \frac{x - x'}{|x - x'|^3} \, d^3x' \quad (3.7.11)$$

provided $h^*(x)$ equals σ^* whenever x is in $\hat{\mathscr{V}}$. The vector q in (3.7.11) is the flux which would result from the distributed source $h^*(x)$, and may therefore be expressed as a gradient of a concentration satisfying the diffusion equation (3.5.1) throughout the region $\hat{\mathscr{V}}$. In general this concentration will not, however, vanish on $\partial \hat{\mathscr{V}}$, so that (3.7.11) does not give the true flux distribution.

We select an $h^*(x)$ of the form

$$h^*(x) = \{\sigma^*/[A_q]\} \sum_i q^*(x_i - x), \quad (3.7.12)$$

where x_i is the position of the ith quencher molecule,

$$q^*(\rho_i) = \begin{cases} 3(4\pi R^{*3})^{-1} - \delta(\rho_i), & \rho_i < R^*, \\ 0 & , & \rho_i > R^*, \end{cases} \quad (3.7.13)$$

the vector $\rho_i = x_i - x$, and $\delta(\rho_i)$ is the three dimensional Dirac delta function. The distribution given by (3.7.12) corresponds to a set of point sinks at positions x_i, each at the center of a spherical region of radius R^* containing a uniformly distributed source. The source and sinks balance so

$$\int_{\mathscr{V}} q^*(\rho_i) \, d^3\rho_i = 0 \quad (3.7.14)$$

Strictly speaking the requirement $h^*(x) = \sigma^*$ in the void is true only in the limit $R^* \to \infty$, but it is more convenient to defer passage to the limit until we have evaluated Ξ and taken the limit $V \to \infty$.

$$\Xi = (16\pi^2 DV)^{-1} \int_{\hat{\mathscr{V}}} \int_{\mathscr{V}} \int_{\mathscr{V}} h^*(x') h^*(x'') \frac{(x - x')}{|x - x'|^3} \cdot \frac{(x - x'')}{|x - x''|^3}$$
$$d^3x \, d^3x' \, d^3x'' \quad (3.7.15)$$

$$= \frac{\sigma^{*2}}{16\pi^2 D [A_q]^2 V} \int_{\hat{\mathscr{V}}} \int_{\mathscr{V}} \int_{\mathscr{V}} \sum_{ij} q^*(x_i - x') q^*(x_j - x'') \frac{(x - x')}{|x - x'|^3} \cdot \frac{(x - x'')}{|x - x''|^3}$$
$$d^3x \, d^3x' \, d^3x'' \quad (3.7.16)$$

We perform first the double sum and the integration over $\hat{\mathscr{V}}$ with respect to x. In adding up the contributions of the various quencher particles, we may replace the summation by an integration, but in doing so we must not forget to distinguish between two types of terms involved, namely, those for which $i = j$, and those for which i and j are different. Transforming to new variables

$$\rho = x' - x$$
$$\rho' = x'' - x$$
$$\rho_i = x_i - x$$

and letting $V \to \infty$, we have

$$
\Xi = \frac{\sigma^{*2}\hat{V}}{16\pi^2 D [A_q]^2 V} \left[[A_q] \iiint f^{(1)}(\rho_1) q^*(\rho_1 - \rho) q^*(\rho_1 - \rho') \right.
$$

$$
\times \frac{\rho \cdot \rho'}{\rho^3 \rho'^3} d^3\rho_1 \, d^3\rho \, d^3\rho'
$$

$$
+ [A_q]^2 \iiiint f^{(2)}(\rho_1, \rho_2) q^*(\rho_1 - \rho) q^*(\rho_2 - \rho')
$$

$$
\left. \times \frac{\rho \cdot \rho'}{\rho^3 \rho'^3} d^3\rho_1 \, d^3\rho_2 \, d^3\rho \, d^3\rho' \right] .
\qquad (3.7.17)
$$

In this equation $[A_q]$ is the concentration of quencher particles, $[A_q] f^{(1)}(\rho_1)$ is the mean concentration of A_q at the point $x + \rho_1$ if the point x is known to be in $\hat{\mathscr{V}}$, and $[A_q]^2 f^{(2)}(\rho_1, \rho_2) d^3\rho_1 \, d^3\rho_2$ is the mean number of quencher pairs with the two different quencher particles located respectively in the volume elements $d^3\rho_1$ and $d^3\rho_2$, at positions ρ_1 and ρ_2, relative to a randomly selected point x in $\hat{\mathscr{V}}$. The integrations in (3.7.17) are over all space. Finally, we allow R^* to become infinite and substitute the result into (3.7.10) to obtain

$$
k_e > \frac{4\pi [A_q] D V / \hat{V}}{\displaystyle\int_a^\infty \rho_1^{-2} f^{(1)}(\rho_1) d\rho_1 + 2\pi [A_q] \int_a^\infty \int_a^\infty \int_{-1}^1 f^{(2)}(\zeta_p, \rho_1, \rho_2) \zeta_p \, d\rho_1 \, d\rho_2 \, d\zeta_p}
$$

$$(3.7.18)$$

Note that in writing (3.7.18) we have taken into account that $f^{(1)}$ depends only on the magnitude ρ_1 of the vector ρ_1, and $\zeta_p \equiv \rho_1 \cdot \rho_2 / \rho_1 \rho_2$. The lower limits on the integrals in ρ_1 and ρ_2 stem from the fact that $f^{(1)}$ vanishes for $\rho_1 < a$ while $f^{(2)}$ vanishes if $\rho_1 < a$ or $\rho_2 < a$, since no A_q particle can be at a distance less than a from any point in $\hat{\mathscr{V}}$.

3.8. Random Quencher Particles

To evaluate the rigorous upper and lower bounds on the effective rate constant given by (3.6.19) and (3.7.18) we must have the information on the distribution of A_q particles contained in the functions $P(\varepsilon)$, $f^{(1)}$, and $f^{(2)}$. For concentrated solutions, obtaining these functions is a difficult problem in statistical mechanics. However, if we assume that there are no forces acting between the quencher particles, then the presence of an A_q molecule at a certain point has no effect on the probability of finding an A_q molecule nearby, even within the reaction zone of the first molecule, so that $f^{(1)}$ and $f^{(2)}$ are unity for ρ_1 and ρ_2 greater than a.

Furthermore, the probability of finding a region of volume v, placed at random in solution, completely empty of A_q particles is $\exp(-v[A_q])$. The volume fraction of solution lying outside the reaction zone is thus

$$\hat{V}/V = \exp\left(-\tfrac{4}{3}\pi a^3 [A_q]\right). \tag{3.8.1}$$

The probability, $P(\varepsilon)\,d\varepsilon$, that a randomly chosen point in solution will lie in the void $\hat{\mathscr{V}}$ with a nearest point on $\partial\hat{\mathscr{V}}$, the reaction zone interface, a distance ε to $\varepsilon + d\varepsilon$ away, is the product of the probability of finding no quencher particle within a spherical volume $4\pi(a+\varepsilon)^3/3$ and the probability of at least one quencher particle within a shell of radii $(\varepsilon + a)$ to $(\varepsilon + a) + d\varepsilon$. The following form of $P(\varepsilon)$ was derived in section 1.5 for a random bed of spheres of density $n(=[A_q])$, we have

$$P(\varepsilon)\,d\varepsilon = 4\pi(\varepsilon + a)^2 [A_q] \exp\left\{-4\pi[A_q](\varepsilon + a)^3/3\right\} d\varepsilon \tag{3.8.2}$$

We now substitute P from (3.8.2) into the inequality (3.6.19) and perform the required integrations, then

$$k_e < (3\gamma_q Da^{-2})\left\{e^{-\gamma_q} - \gamma_q^{1/3}\psi_{2/3}(\gamma_q)\right\}^{-1}$$
$$\gamma_q = 4\pi[A_q]a^3/3 \tag{3.8.3}$$
$$\psi_{2/3}(\gamma_q) = \int_{\gamma_q}^{\infty} y^{-1/3} e^{-y}\,dy$$

where $\psi_{2/3}(\gamma_q)$ is the incomplete Gamma function (Abramowitz and Stegun, 1964). Examination of the bound represented by (3.8.3) for very dilute suspensions gives us

$$k_e < 4\pi Da[A_q] \qquad [A_q] \text{ small} \tag{3.8.4}$$

For very dilute suspensions, the true behavior of k_e is well-known (Smoluchowski 1915; 1917). At low concentrations different A_q behave independently of one another, each particle acting as if it were alone in an infinite volume of fluid. Furthermore, σ^* can be neglected in (3.5.1) as it depends linearly on $[A_q]$, and it is straightforward to show that (3.8.4) is, in fact, an equality. At higher concentrations the quencher particles do not act independently in that the diffusion of B^* to one quencher will be affected by the presence of other quencher particles. As the quencher concentration $[A_q]$ is raised, the interaction between quencher particles becomes increasingly important, k_e deviates more and more from the linear form predicted by (3.8.4). For very concentrated suspensions (3.8.3) becomes

$$\lim_{[A_q] \to \infty} k_e < 16\pi^2 D a^4 [A_q]^2 \exp\left\{4\pi[A_q]a^3/3\right\} \tag{3.8.5}$$

Substituting $f^{(1)} = f^{(2)} = 1$ into inequality (3.7.18) we obtain for the lower bound on k_e

$$k_e > 4\pi D a [A_q] \exp\left\{4\pi[A_q]a^3/3\right\} \tag{3.8.6}$$

This result is also seen to be exact for the limit of infinite dilution. Indeed the exponential by which the right-hand side of (3.8.6) differs from the infinite dilution result (3.8.4) has a quite trivial significance; it merely takes into account those B^* molecules, which are generated inside the reaction zone and therefore quenched instantly.

The upper and lower bounds given by (3.8.3) and (3.8.6) are plotted in Fig. 3.8.1 (top and bottom curves). We see that they diverge quite rapidly, suggesting that the comparatively simple trial functions which we have used require a certain amount of elaboration, with a consequent increase in the information needed concerning the distribution of quencher particles in solution.

An improvement in the lower bound can be obtained relatively simply by using, in place of the trial source distribution (3.7.12) an $h^*(x)$ of the form

$$h^*(x) = \begin{cases} \sigma^* & , \quad x \text{ in } \hat{\mathscr{V}} \\ -\sigma^* \hat{V}/(V - \hat{V}), & x \text{ in } \mathscr{V} - \hat{\mathscr{V}} \end{cases} \tag{3.8.7}$$

This satisfies condition (3.7.2), as well as the requirement that the mean value of h^* over the entire system be zero. Substitution from (3.8.7) into (3.7.10) leads, after some reduction, to the inequality

$$k_e > \frac{16\pi^2 D(V - \hat{V})^2/V^2}{\displaystyle\iint G(\boldsymbol{\rho}, \boldsymbol{\rho}')(\boldsymbol{\rho} \cdot \boldsymbol{\rho}'/\rho^3 \rho'^3)\, d^3\boldsymbol{\rho}\, d^3\boldsymbol{\rho}'}, \tag{3.8.8}$$

where $G(\boldsymbol{\rho}, \boldsymbol{\rho}')$ is the probability that, if x is a randomly chosen point in $\mathscr{V}$, both it and the pair of points $x + \boldsymbol{\rho}$ and $x + \boldsymbol{\rho}'$ will all three be found to be in $\mathscr{V}$.

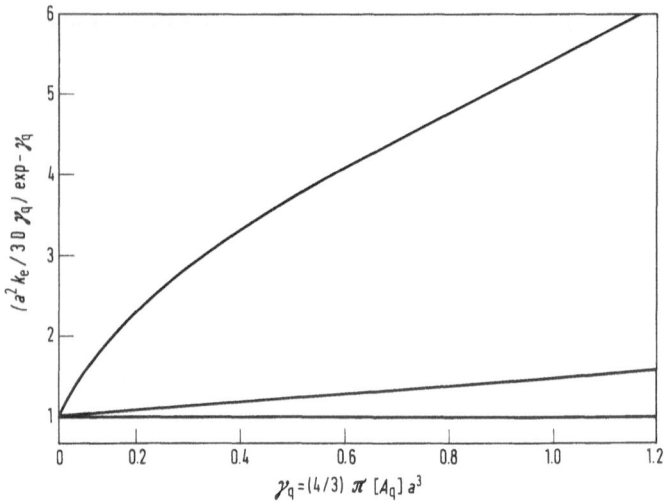

Fig. 3.8.1. Bounds on the reduced $k_e[A_q]^{-1}$ ratio as functions of the reduced quencher concentration. The top curve is the upper bound (3.8.3), the bottom line is the lower bound (3.8.6), and the middle curve is the improved lower bound (3.8.8).

The probability $G(\rho, \rho')$ is discussed in section 1.3, and its form for spherical quencher particles placed at random, independently of one another, is given by (1.3.11)–(1.3.14). These equations for $G(\rho, \rho')$ permit a computer calculation of the formula in (3.8.8). The lower bound obtained in this way is represented by the middle curve in Fig. 3.8.1.

Chapter 4

Heterogeneous Catalysis

4.1. Introduction

The phenomenon of diffusion and reaction in catalyst pellets has engaged the attention of chemical engineers for the past thirty-five years (Thiele, 1967), though the basic problem was solved many years before and lay neglected (Jüttner 1909; see notes on the history of the subject in Ch. 1 of Aris, 1974). The calculation of effectiveness factors for various reaction kinetics, pellet geometries, and models of the diffusion-reaction process provides a number of interesting problems that may be approached using variational methods. We shall use complementary variational principles to obtain both upper and lower bounds on the effectiveness factor in a homogeneous slab of catalyst. A measure of the accuracy achieved by the variational calculation is the closeness of these bounds. Calculations will be carried out both for linear and for Langmuir-Hinshelwood kinetics. Some general results concerning pellet shape can also be obtained using variational principles, and it is shown in section 4.3 that of all catalyst pellets having a given volume the spherical pellet has the least effectiveness factor when the kinetics are of the first order.

The homogeneous model for a heterogeneous catalyst is not entirely suitable when we consider some of the zeolite or molecular sieve catalysts in use today. Here, the porous medium is relatively inert and the reaction takes place on or in a number of finer particles, distributed throughout the pellet. Variational bounds are calculated on the effectiveness factor for a first order irreversible reaction that occurs on discrete surfaces distributed throughout a slab of catalyst. The use of complementary variational principles to analyse experimental data is also discussed.

4.2. Variational Principles for Heterogeneous Catalysis: the Homogeneous Model

The most commonly used model for diffusion and reaction is the homogeneous model, in which it is assumed that the complexities of diffusion in a porous medium can be reduced to a Fickian diffusion with an effective diffusion coefficient. The estimation of such effective diffusion coefficients in a porous medium was discussed at some length in Chapter 2. It is also assumed that the reactive surface is finely dispersed, so that the reaction takes place everywhere. If c is the concentration of the species disappearing by reaction at a rate $r(c)$ per unit volume of catalyst, and D_e is the effective diffusion coefficient, then

$$\nabla \cdot (D_e \nabla c) = r(c) \qquad (4.2.1)$$

at any point x within the volume $\mathscr{V}$ of the catalyst pellet. We can without loss of generality assume that $r(0) = 0$. If there is mass transfer resistance at the surface of the pellet, $\partial \mathscr{V}$, the solution of (4.2.1) must satisfy the boundary condition

$$D_e \boldsymbol{n} \cdot \nabla c = k_c(c_f - c) \qquad (4.2.2)$$

where k_c is a mass transfer coefficient, c_f is the reactant concentration far from the catalyst pellet and $\boldsymbol{n}$ is the outward unit normal vector for the external surface $\partial \mathscr{V}$ of the catalyst pellet. Note that $k_c(c_f - c)$ from (4.2.2) must remain finite for any k_c, so that in the limit of perfect external mass transfer ($k_c \to \infty$) the external boundary condition becomes

$$c = c_f \qquad (4.2.3)$$

A number of dimensionless quantities will now be introduced. The effectiveness factor η is defined as the ratio of the rate of reaction within the catalyst to the rate of reaction within the catalyst with no diffusion restriction,

$$\eta = \int_{\mathscr{V}} r(c) \, d^3\boldsymbol{x} / Vr(c_f). \qquad (4.2.4)$$

It can be rewritten with the aid of (4.2.1) and the divergence theorem in terms of an integration over the external surface $\partial \mathscr{V}$ of the catalyst pellet

$$\eta = \int_{\partial \mathscr{V}} D_e \boldsymbol{n} \cdot \nabla c \, d^2\boldsymbol{x} / Vr_f \qquad (4.2.5)$$

where $r_f = r(c_f)$. In addition let $L/2$ be the ratio of the total pellet volume to external surface area and define the Thiele modulus Λ by

$$\Lambda^2 = L^2 r_f / 4D_e c_f \tag{4.2.6}$$

and

$$N = Lk_c / 2D_e. \tag{4.2.7}$$

4.2a. Upper Bound

Upper and lower bound variational principles for this problem were first formulated by Arthurs (1969). The solution of the differential equation (4.2.1) subject to the boundary condition (4.2.2) is entirely equivalent to minimizing* the functional $\mathscr{J}$

$$Vc_f r_f \mathscr{J}(u) \equiv \int_{\mathscr{V}} \left(D_e \nabla u \cdot \nabla u + 2F(u)\right) d^3 x$$
$$+ \int_{\partial \mathscr{V}} k_c (c_f - u)^2 d^2 x \tag{4.2.8}$$

with

$$F(u) = \int_0^u r(\zeta)\, d\zeta$$

The trial concentration function u must be continuous, have continuous first derivatives and at least piece-wise continuous second derivatives in $\mathscr{V}$, and satisfy

$$k_c |c_f - u| < \infty \quad \text{on } \partial \mathscr{V}. \tag{4.2.9}$$

We consider the variations u_1 around u_0 and substitute $u = u_0 + u_1$ into (4.2.8). The terms in $\mathscr{J}$ which are first order in the variation u_1 are

$$Vc_f r_f \mathscr{J}_1 = \int_{\mathscr{V}} \{2D_e \nabla u_1 \cdot \nabla u_0 + 2u_1 r(u_0)\} d^3 x$$
$$- 2 \int_{\partial \mathscr{V}} u_1 k_c (c_f - u_0) d^2 x \tag{4.2.10}$$

---

*The reader can verify that

$$\mathscr{H}(\mathbf{q}, u) = \frac{\mathbf{q} \cdot \mathbf{q}}{2D_e} - \int_0^u r(\zeta)\, d\zeta \quad \text{in } \mathscr{V}$$

and

$$\mathscr{A}(\mathbf{q}, u) = c_f (\mathbf{n} \cdot \mathbf{q}) - \frac{(\mathbf{n} \cdot \mathbf{q})^2}{2k_c} \quad \text{on } \partial \mathscr{V}$$

for the variational functional $\mathscr{F}(\mathbf{q}, u)$ defined by (1.6.2), along with the scalar selection procedure (1.6.13) does indeed give the variational upper bound principle (4.2.8). Furthermore the $\mathscr{H}(\mathbf{q}, u)$ and $\mathscr{A}(\mathbf{q}, u)$ forms combined with the vector field selection procedure (1.6.17) also give after a trivial integration by parts, the lower bound variational principle (4.2.15).

$$= 2 \int_{\mathscr{V}} u_1 \{ - \nabla \cdot (D_e \nabla u_0) + r(u_0) \} \, d^3x$$

$$+ 2 \int_{\partial \mathscr{V}} u_1 \{ D_e \mathbf{n} \cdot \nabla u_0 - k_c(c_f - u_0) \} \, d^2x \qquad (4.2.11)$$

where the second expression follows from the application of Gauss' theorem. The Euler-Lagrange equations are exactly (4.2.1) and (4.2.2), and $\mathscr{J}_1$ vanishes if and only if u_0 is taken to be the reactant concentration. After some rearrangement the variational expression (4.2.8) becomes

$$\mathscr{J}(u) = \mathscr{J}(c) + \mathscr{J}_2 \qquad (4.2.12)$$

where the second variation is

$$V c_f r_f \mathscr{J}_2 = \int_{\mathscr{V}} \left[D_e (\nabla u_1)^2 + (u_1)^2 \, r'(c + \theta_m u_1) \right] d^3x$$

$$+ \int_{\partial \mathscr{V}} k_c (u_1)^2 \, d^2x \qquad (4.2.13)$$

with $0 < \theta_m < 1$. Now if the reaction rate is monotonic increasing, r' is positive and so is $\mathscr{J}_2$. Thus the upper bound given by the variational principle

$$\mathscr{J}(u) \geq \mathscr{J}(c) \qquad (4.2.14)$$

4.2b. Lower bound

A lower bound variational principle can be formulated starting from the functional $\mathscr{G}(\mathbf{q})$

$$V c_f r_f \mathscr{G}(\mathbf{q}) = - \int_{\mathscr{V}} \left[\frac{\mathbf{q} \cdot \mathbf{q}}{D_e} + 2I(\nabla \cdot \mathbf{q}) \right] d^3x$$

$$+ \int_{\partial \mathscr{V}} \left[2c_f \mathbf{n} \cdot \mathbf{q} - \frac{(\mathbf{n} \cdot \mathbf{q})^2}{k_c} \right] d^2x \qquad (4.2.15)$$

with

$$I(\nabla \cdot \mathbf{q}) = \int_0^{(\nabla \cdot \mathbf{q})} r^{-1}(\zeta) \, d\zeta \qquad (4.2.16)$$

where r^{-1} is the inverse of the reaction rate function (4.2.1). (See the preceding footnote). The trial flux $\mathbf{q}$ for $\mathscr{G}$ must be continuous with at least piecewise continuous first derivatives.

The trial flux vector $\mathbf{q}_0 + \mathbf{q}_1$ is now introduced into (4.2.15) and the terms in $\mathscr{G}$ which are of the first order in the variation $\mathbf{q}_1$ are

$$Vc_f r_f \mathscr{G}_1 = - \int_{\mathscr{V}} \left[2 \frac{\boldsymbol{q}_0 \cdot \boldsymbol{q}_1}{D_e} + 2(\nabla \cdot \boldsymbol{q}_1) r^{-1} (\nabla \cdot \boldsymbol{q}_0) \right] d^3 \boldsymbol{x}$$

$$+ \int_{\partial \mathscr{V}} \left[2c_f \boldsymbol{n} \cdot \boldsymbol{q}_1 - 2 \frac{(\boldsymbol{n} \cdot \boldsymbol{q}_0)(\boldsymbol{n} \cdot \boldsymbol{q}_1)}{k_c} \right] d^2 \boldsymbol{x}$$

$$= 2 \int_{\mathscr{V}} \boldsymbol{q}_1 \cdot \left[- \frac{\boldsymbol{q}_0}{D_e} + \nabla r^{-1} (\nabla \cdot \boldsymbol{q}_0) \right] d^3 \boldsymbol{x}$$

$$+ 2 \int_{\partial \mathscr{V}} (\boldsymbol{n} \cdot \boldsymbol{q}_1) \left[c_f - r^{-1} (\nabla \cdot \boldsymbol{q}_0) - \frac{(\boldsymbol{n} \cdot \boldsymbol{q}_0)}{k_c} \right] d^2 \boldsymbol{x} \quad (4.2.17)$$

Again the second expression results from the application of Gauss' theorem. The Euler-Lagrange equations are

$$\boldsymbol{q}_0 - D_e \nabla r^{-1} (\nabla \cdot \boldsymbol{q}_0) = 0 \quad (4.2.18)$$

and

$$k_c \left[c_f - r^{-1} (\nabla \cdot \boldsymbol{q}_0) \right] - \boldsymbol{n} \cdot \boldsymbol{q}_0 = 0. \quad (4.2.19)$$

We note from (4.2.18) that $\boldsymbol{q}_0$ is D_e times the gradient of some scalar function γ_c, where $\gamma_c = r^{-1} (\nabla \cdot \boldsymbol{q}_0)$ and that

$$\nabla \cdot \boldsymbol{q}_0 = r(\gamma_c) \quad (4.2.20)$$

The Euler-Lagrange equations are completely equivalent to Eqs. (4.2.1) and (4.2.2), and $\mathscr{G}_1$ vanishes if γ_c is taken to be the reactant concentration. The variational principle $\mathscr{G}$ can now be written

$$\mathscr{G}(\boldsymbol{q}) = \mathscr{G}(D_e \nabla c) + \mathscr{G}_2$$

where the second variation is

$$Vc_f r_f \mathscr{G}_2 = - \int_{\mathscr{V}} \left[\frac{\boldsymbol{q}_1 \cdot \boldsymbol{q}_1}{D_e} + (r^{-1})' \left(r(c) + \theta_m \nabla \cdot \boldsymbol{q}_1 \right) (\nabla \cdot \boldsymbol{q}_1)^2 \right] d^3 \boldsymbol{x}$$

$$- \int_{\partial \mathscr{V}} \frac{(\boldsymbol{n} \cdot \boldsymbol{q}_1)^2}{k_c} d^2 \boldsymbol{x}, \qquad 0 < \theta_m < 1, \quad (4.2.21)$$

and $(r^{-1})'$ denotes the derivative of the inverse function r^{-1}. This is clearly positive for any monotonic increasing r, since evaluated in (4.2.21) at $\left(r(c) + \theta_m \nabla \cdot \boldsymbol{q}_1 \right)$

$$dr^{-1}(\zeta)/d\zeta = 1/(dr(v)/dv) > 0 \quad (4.2.22)$$

when $\zeta = r(v)$. Thus $\mathscr{G}_2$ is negative and the lower bound

$$\mathscr{G}(D_e \nabla c) \geq \mathscr{G}(\boldsymbol{q}) \quad (4.2.23)$$

is established.

To equate maximum $\mathcal{G}$ to the minimum of $\mathcal{J}$ we first note the relation of F and I. Upon substitution of $\zeta = r(v)$ we see that F from (4.2.8) and I from (4.2.16) are related by

$$I[r(c)] = \int_0^{r(c)} r^{-1}(\zeta)\, d\zeta = cr(c) - F(c) \tag{4.2.24}$$

The maximum $\mathcal{G}$ is evaluated at $\boldsymbol{q} = D_e \nabla c$, and with the aid of (4.2.1) and (4.2.24) the maximum becomes

$$Vc_f r_f \mathcal{G}_{\max} = -\int_{\mathscr{V}} \left[D_e \nabla c \cdot \nabla c + 2cr(c) - 2F(c) \right] d^3 \boldsymbol{x}$$
$$+ \int_{\partial \mathscr{V}} D_e (\boldsymbol{n} \cdot \nabla c) \left[2c_f - \left(\frac{D_e}{k_c} \right) (\boldsymbol{n} \cdot \nabla c) \right] d^2 \boldsymbol{x} \tag{4.2.25}$$

By using Gauss' theorem as well as the differential equation (4.2.1) and the boundary conditions (4.2.2) to rearrange the first term of the volume integration in (4.2.25) we find

$$Vc_f r_f \mathcal{G}_{\max} = -\int_{\mathscr{V}} \left(cr(c) - 2F(c) \right) d^3 \boldsymbol{x}$$
$$+ c_f \int_{\partial \mathscr{V}} D_e (\boldsymbol{n} \cdot \nabla c)\, d^2 \boldsymbol{x} \tag{4.2.26}$$

If the differential equation (4.2.1) is substituted into the first term in the volume integral of (4.2.26), Gauss' theorem is again applied, and the boundary condition (4.2.2) is substituted into the surface terms to eliminate $D_e \boldsymbol{n} \cdot \nabla c$, we obtain

$$Vc_f r_f \mathcal{G}_{\max} = \int_{\mathscr{V}} \left[D_e (\nabla c \cdot \nabla c) + 2F(c) \right] d^3 \boldsymbol{x}$$
$$+ \int_{\partial \mathscr{V}} k_c (c_f - c)^2\, d^2 \boldsymbol{x}$$
$$= Vc_f r_f \mathcal{J}(c) = Vc_f r_f \mathcal{J}_{\min} \tag{4.2.27}$$

and the maximum of $\mathcal{G}$ is equal to the minimum of $\mathcal{J}$. Using the definition of the effectiveness factor given by (4.2.5) we can rewrite (4.2.26) and (4.2.27) in the form

$$\mathcal{G}_{\max} = \mathcal{J}_{\min} = \eta - \frac{1}{c_f r_f V} \int_{\mathscr{V}} R(c)\, d^3 \boldsymbol{x} = \eta - M \tag{4.2.28}$$

where

$$R(c) = cr(c) - 2F(c). \tag{4.2.29}$$

$R(c)$ vanishes only if $r(c)$ is linear, but it can often be bounded. For example with Langmuir-Hinshelwood kinetics $r(c) = k'_r c/(1 + Kc)$ the value of M lies between zero and

$$1 - \frac{2(1 + \kappa)\left[\kappa - \ln(1 + \kappa)\right]}{\kappa^2} = -\frac{\kappa}{3} + \frac{\kappa^2}{6} \cdots$$

where

$$\kappa = Kc_f.$$

4.3. Application to First Order Kinetics in a Slab

To illustrate the application of these bounds Rester and Aris (1972) treated the simplest diffusion and reaction problem, a first order isothermal reaction occurring in an isotropic porous flat slab of thickness L. Let x be the distance from the center plane to a given point, and $\mathbf{i}_x$ be a unit vector in the positive x-direction. Assuming the diffusion coefficient D_e is a constant, we have for the reactant concentration c

$$D_e \frac{d^2 c}{dx^2} = k_r c \quad \text{in } -L/2 < x < L/2 \qquad (4.3.1)$$

with the boundary conditions

$$\pm D_e \mathbf{i}_x \cdot \nabla c = k_c(c_f - c) \quad \text{at } x = \pm L/2. \qquad (4.3.2)$$

The mass transfer coefficient k_c and first order rate constant are also assumed to be concentration independent. The variational principles (4.2.8) and (4.2.15) provide upper and lower bounds on the effectiveness factor η

$$\mathscr{G}(\mathbf{q}) \leq \eta \leq \mathscr{J}(u). \qquad (4.3.3)$$

We can easily obtain the forms of $\mathscr{J}$ and $\mathscr{G}$ for a one-dimensional slab from (4.2.8) and (4.2.15) respectively by replacing V by the slab length L, reducing the volume integration $d^3 x$ to an integration over the length coordinate x from $-L/2$ to $L/2$, and by dropping the surface integral in order to evaluate its integrand at $x = -L/2$ and $x = L/2$.

The reaction rate written in terms of the concentration c_f far from the particle is $r_f = k_r c_f$. There are two parameters, the Thiele modulus

$$\Lambda^2 = L^2 k_r/4D_e \qquad (4.3.4)$$

and

$$N = Lk_c/2D_e \qquad (4.3.5)$$

4.3a. Variational Estimates for Small Λ

For small Λ the concentration profile is fairly flat and we expect a trial function of the form

$$u = c_f \chi [\xi + (1 - \xi) \tilde{x}^2] \qquad (4.3.6)$$

will be suitable, where $\tilde{x} = 2x/L$ is the dimensionless distance from the center plane of the slab. For the lower bound a suitable trial function is

$$\boldsymbol{q} = 2D_e \tilde{x} u \boldsymbol{i}_x / L. \qquad (4.3.7)$$

The procedure is to substitute the trial functions into the bounds, then minimize the difference between the results by choosing the constants χ and ξ. The trial functions (4.3.6) and (4.3.7) are substituted into the variational principles (4.2.8) and (4.2.15) respectively, the algebraic steps are straightforward and we obtain

$$\frac{\Lambda^2}{N} + \frac{1 + (2\Lambda^2/5)}{1 + (\Lambda^2/15)} < \frac{1}{\eta} < \frac{\Lambda^2}{N} + \frac{1 + (3\Lambda^2/7) + (\Lambda^4/105)}{1 + (2\Lambda^2/21)} \qquad (4.3.8)$$

The bounds are expressed in terms of the dimensionless quantities Λ and N defined by (4.3.4) and (4.3.5). The curves on the left-hand side of Fig. 4.3.1 show these bounds as a percentage error from the true relation.

$$\frac{1}{\eta} = \frac{\Lambda^2}{N} + \frac{\Lambda}{\tanh \Lambda} \qquad (4.3.9)$$

As might be expected the bounds diverge as Λ increases since the trial functions become less suitable for larger Λ.

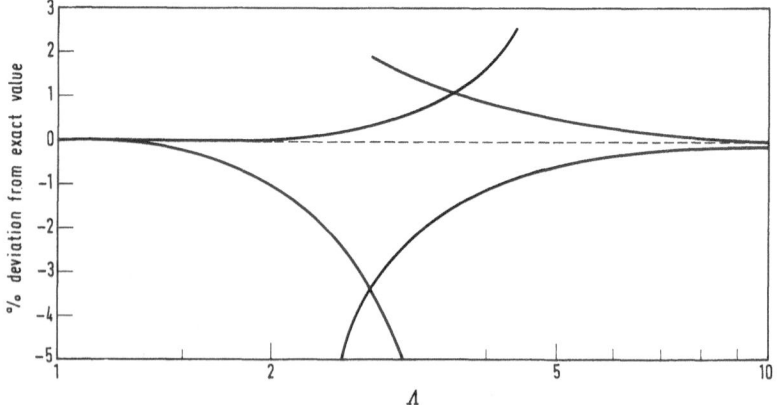

Fig. 4.3.1. Example of bounds for homogeneous problem.

4.3b. *Variational Estimates for Large Λ*

For large Λ the reaction occurs near the surface, and the concentration profile falls off rapidly. Thus, trial functions of the form

$$u = \begin{cases} c_f \chi \left\{ \dfrac{\tilde{x} - \beta}{1 - \beta} \right\}^b & \beta < \tilde{x} < 1 \\[2ex] 0 & -\beta < \tilde{x} < \beta \\[2ex] c_f \chi \left\{ \dfrac{\tilde{x} + \beta}{-1 + \beta} \right\}^b & -1 < \tilde{x} < -\beta \end{cases}$$

with $b > 2$ and   (4.3.10)

$$q = 2D_e u i_x / L \tag{4.3.11}$$

should be appropriate trial functions for larger λ. The trial function (4.3.10) is substituted into the upper bound (4.2.8), the lower bound trial function (4.3.11) is introduced into (4.2.15), the resulting bounds are expressed in terms of Λ, N and the variational parameters χ, β, and b. Note that (4.3.10) gives an even trial function for the upper bound calculation; in order to obtain a non-negative lower bound on η the trial flux (4.3.11) must be an odd function. The resulting bounds on the effectiveness factor η are

$$\chi^2 g_\beta + N \Lambda^{-2} (1 - \chi)^2 > \eta > 2\chi \Lambda^{-2} - \chi^2 N^{-1} \Lambda^{-2} - \chi^2 \Lambda^{-2} g_\beta \tag{4.3.12}$$

where

$$g_\beta (\beta, b) = \frac{b^2}{\Lambda^2 (2b - 1)(1 - \beta)} + \frac{(1 - \beta)}{(2b + 1)} \tag{4.3.13}$$

Clearly the best choice of β for both bounds is the value that minimizes g_β namely

$$1 - \frac{b}{\Lambda} \sqrt{\frac{2b + 1}{2b - 1}} \tag{4.3.14}$$

But β must never be negative, so that b must be chosen to satisfy

$$\Lambda \geq b \sqrt{\frac{2b + 1}{2b - 1}} \tag{4.3.15}$$

Selecting the extremum values of χ is a matter of straightforward algebra, and substituting the extremizing values of β and χ into the upper and lower bounds (4.3.8), gives

$$\frac{\Lambda^2}{N} + \Lambda \sqrt{1 - \frac{1}{4b^2}} < \frac{1}{\eta} < \frac{\Lambda^2}{N} + \Lambda \bigg/ \sqrt{1 - \frac{1}{4b^2}} \qquad (4.3.16)$$

It appears from (4.3.16) that the best possible bounds are obtained for the largest permissible value of b. This occurs when the inequality restriction (4.3.15) is taken to be an equality. As Λ increases these bounds can be written with considerable accuracy in the form

$$\frac{\Lambda^2}{N} + \Lambda - \frac{1}{8\Lambda} - \frac{1}{8\Lambda^2} + 0\left(\frac{1}{\Lambda^3}\right) < \frac{1}{\eta} < \frac{\Lambda^2}{N} + \Lambda + \frac{1}{8\Lambda} + \frac{1}{8\Lambda^2} + 0\left(\frac{1}{\Lambda^3}\right)$$
$$(4.3.17)$$

The curves to the right in Fig. 4.3.1 show these bounds; as is to be expected, they diverge for smaller values of Λ.

4.4. Application to Langmuir-Hinshelwood Kinetics in a Slab

The linear example in the previous paragraph is a convenient one with which to introduce the subject as its known solution provides a useful test of the accuracy of the bounds. A more exacting and valuable application however is to nonlinear kinetics for here there is no analytic solution and the numerical solution can often be a laborious affair. A measure of the efficacy of the variational method will be the closeness with which the calculated upper and lower bounds approach one another. Langmuir-Hinshelwood kinetics provides a nonlinear example of interest and importance.

An irreversible reaction

$$A_1 + a_2 A_2 + \ldots \rightarrow b_1 B_1 + b_2 B_2 + \ldots \qquad (4.4.1)$$

takes place without any change in the number of moles on the internal surface of a porous catalyst slab. The rate at which A_1 disappears depends on the local concentrations of all the chemical species

$$r = k_r c_{A_1} \bigg/ \left(1 + \sum_i K_{A_i} c_{A_i} + \sum_i K_{B_i} c_{B_i}\right) \qquad (4.4.2)$$

where c_{A_i} and c_{B_i} $(i = 1, 2, \ldots)$ denotes the concentrations of reactants and products respectively. The general derivation of a kinetic rate expression for (4.2.1) will not be discussed in detail here; it will suffice to mention several important cases in which the reaction rate expression can be made a function of c_{A_1} (or c) alone as is required by Eq. (4.2.1). In a binary system when the reaction is $A \rightarrow B$, the total pressure and concentration are constant so that $c_B = c_T - c_A$. Then

$$r = \frac{k_r c_A}{1 + K_A c_A + K_B c_B} = \frac{k'_r c_A}{1 + K c_A} \qquad (4.4.3)$$

where

$$k_r' = k_r/(1 + K_B c_T) \tag{4.4.4}$$

and

$$K = (K_A - K_B)/(1 + K_B c_T). \tag{4.4.5}$$

For a catalyst with very fine pores within which the so-called Knudsen regime of diffusion prevails, and when the external mass transfer resistance is negligible ($k_c \to \infty$), Satterfield and Roberts (1965) have shown that (4.4.3) can be used in terms of c_{A_1} with

$$k_r' = k_r/\omega \tag{4.4.6}$$

$$K = \left[(K_{A_1}/D_{eA_1}) + a_2(K_{A_2}/D_{eA_2}) \ldots - b_1(K_{B_1}/D_{eB_1}) - \ldots \right] D_{eA_1}/\omega \tag{4.4.7}$$

where

$$\omega = \left[1 + K_{A_2}\{c_{A_2 f} - a_2 c_{A_1 f}(D_{eA_1}/D_{eA_2})\} + \ldots \right.$$
$$\left. + K_{B_1}\{c_{B_1 f} + b_1 c_{A_1 f}(D_{eA_1}/D_{eB_1})\} + \ldots \right] \tag{4.4.8}$$

and $c_{A_{if}}$ and $c_{B_{if}}$ are concentrations of A_i and B_i far from the catalyst pellet. We have assumed, as is customary in these calculations, that all the effective diffusion coefficients D_{eA_i}, D_{eB_i} are a constant, and independent of reactant or product concentration (Satterfield, 1970). Though the values of both k_r' and K can be negative, as Satterfield has pointed out, we shall consider only positive k_r' but will permit K to have either sign.

Dropping the subscript A_1 (or A), we note that the kinetic rate expression from (4.4.3)

$$r(c) = k_r' c/(1 + Kc), \tag{4.4.9}$$

and also that its derivative

$$dr/dc = k_r'/(1 + Kc)^2 > 0 \tag{4.4.10}$$

is positive as required for the variational inequalities.

For the steady state in a one-dimensional slab of thickness L the material balance on species A_1 (or A) gives

$$D_e \frac{d^2 c}{dx^2} = \frac{k_r' c}{1 + Kc} \qquad -\frac{L}{2} \le x \le \frac{L}{2} \tag{4.4.11}$$

with the boundary conditions

$$\pm D_e \boldsymbol{i_x} \cdot \nabla c = k_c(c_f - c) \qquad x = \pm L/2 \tag{4.4.12}$$

where x is the distance from the centerplane, $\boldsymbol{i_x}$ is a unit vector in the positive x-direction, and c_f is the reactant concentration far from the slab. The effective diffusion coefficient D_e, the external mass transfer coefficient, and the constants k'_r and K from the rate expression (4.4.9) are all assumed to be concentration independent. We can easily obtain the forms of $\mathscr{J}$ and $\mathscr{G}$ for a one-dimensional slab from (4.2.8) and (4.2.15) respectively by replacing V by the slab length L, by reducing the volume integration d^3x to an integration over the length coordinate x from $-L/2$ to $L/2$, and by dropping the surface integrations over $\partial \mathscr{V}$ entirely in order to evaluate their integrand at $x = -L/2$ and $x = L/2$.

The catalyst geometry is the same as in the previous section, a flat slab of thickness L; we recall that $\tilde{x}$ is the fractional distance from the center plane of the slab to the external surface,

$$N = L k_c / 2 D_e, \tag{4.4.13}$$

and

$$\Lambda^2 = L^2 r(c_f) / 4 D_e c_f. \tag{4.4.14}$$

We note that the upper bound given by (4.2.8) and the lower bound given by (4.2.15) do not bound the effectiveness factor η directly, but rather

$$\mathscr{G} \le \eta - M \le \mathscr{J} \tag{4.4.15}$$

where

$$M = \frac{1}{2 c_f r(c_f)} \int_{-1}^{1} R(c) \, d\tilde{x} \tag{4.4.16}$$

and

$$R(c) = c r(c) - 2 \int_{0}^{c} r(\zeta) \, d\zeta$$

Further we note that $R(c)$ for Langmuir-Hinshelwood kinetics can be bounded at c_f and 0, thus the value of M must lie between zero and

$$1 - \frac{2(1 + \kappa)\{\kappa - \ln(1 + \kappa)\}}{\kappa^2} = -\frac{\kappa}{3} + \frac{\kappa^2}{6} \ldots \tag{4.4.17}$$

where

$$\kappa = K c_f \tag{4.4.18}$$

4.4a. The Variational Estimate of an Upper Bound for Small Λ

For small values of the parameter (roughly $\Lambda < 2$) a suitable trial function is

$$u = c_f \chi [\xi + (1 - \xi) \tilde{x}^2] \tag{4.4.19}$$

where $\tilde{x} (= 2x/L)$ is a dimensionless distance from the centerplane of the slab. The parameter χ is chosen to satisfy the boundary conditions (4.4.12), namely

$$\chi = N / [2(1 - \xi) + N], \tag{4.4.20}$$

and ξ is a disposable parameter, $0 \leq \xi \leq 1$. The trial function (4.4.19) is introduced into the upper bound (4.2.8), and after some straightforward integrations (See 4.4.A.) we obtain the upper bound

$$\mathscr{J}(u) = E - 4(1 + \kappa) Q/\kappa^2 \tag{4.4.21}$$

where

$$\begin{aligned}
E = {} & [4\chi^2(1 - \xi)^2 + 3N(1 - \chi)^2]/(3\Lambda^2) \\
& + \tfrac{2}{3}\chi\kappa^{-1}(1 + \kappa)(1 + 2\xi) \\
& - 2\kappa^{-2}(1 + \kappa)[\ln(1 + \chi\kappa) - 2]
\end{aligned} \tag{4.4.22}$$

and

$$Q_+ = \sqrt{\frac{1 + \chi\xi\kappa}{\chi(1 - \xi)\kappa}} \, \text{arc tan} \sqrt{\frac{\chi(1 - \xi)\kappa}{1 + \chi\xi\kappa}} \quad \text{if } \kappa > 0 \tag{4.4.23}$$

or

$$Q_- = \sqrt{\frac{1 + \chi\xi\kappa}{\chi(\xi - 1)\kappa}} \, \text{arc tanh} \sqrt{\frac{\chi(\xi - 1)\kappa}{1 + \chi\xi\kappa}} \quad \text{if } -1 < \kappa < 0 \tag{4.4.24}$$

The values of ξ which give the smallest upper bound must be found by direct computation. Some typical values are given in Table 4.4.1.

4.4b. The Variational Estimate of a Lower Bound for Small Λ

For the lower bound a suitable trial function is

$$\boldsymbol{q} = 2D_e L^{-1}(du/d\tilde{x})\, \boldsymbol{i}_x = (4D_e c_f/L)\chi(1 - \xi)\tilde{x}\boldsymbol{i}_x \tag{4.4.25}$$

where $\tilde{x}(= 2x/L)$ runs from -1 to 1. Here it is convenient to combine χ and ξ into a single parameter

$$\mu = 2\chi(1 - \xi) \geq 0. \tag{4.4.26}$$

Then the lower bound is

$$\mathscr{G}(\boldsymbol{q}) = 2(1 + \kappa)\kappa^{-1}\{\mu\Lambda^{-2} - \mu^2\mathscr{L} + \kappa^{-1}\ln[1 - \kappa\mu\Lambda^{-2}(1 + \kappa)^{-1}]\} \tag{4.4.27}$$

where

$$\mathscr{L} = \frac{\kappa(N + 3)}{6N\Lambda^2(1 + \kappa)} \tag{4.4.28}$$

Arthurs (1969) has pointed out that constraints may be put on the trial function by the existence of the inverse function r^{-1} when using the variational principle (4.2.15) for the lower bound. Indeed an even stronger restriction may be required for the existence of the integral I

of the lower bound (4.2.16), and in this case we require that

$$\mu < \Lambda^2 (1 + \kappa)/\kappa \quad \text{when } \kappa > 0 \qquad (4.4.29)$$

The best lower bound is given by the value of μ that maximizes $\mathscr{G}$ and this satisfies the quadratic

$$\mu^2 - \mu q_\mu + (2\mathscr{L})^{-1} = 0 \qquad (4.4.30)$$

where

$$q_\mu = \frac{1}{2\mathscr{L}\Lambda^2} + \frac{(1 + \kappa)\Lambda^2}{\kappa} \qquad (4.4.31)$$

The appropriate root to take depends on the sign of κ and is $\mu = \frac{1}{2}q_\mu \mp \frac{1}{2}\{q_\mu^2 - (2/\mathscr{L})\}^{1/2}$ according as $\kappa > 0$ (negative sign) or $-1 < \kappa < 0$ (positive sign). Some typical values of the bounds and the best values of ξ are given in Table 4.4.1. For $\Lambda < 1$ the two bounds are within 2% of one another but as Λ increases the bounds are less close. (See appendix 4.4.2 for a discussion of error bounds on the variational solution u).

Table 4.4.1 Bounds on $(\eta - M)$ for $N = 100$

κ	Λ	$\mathscr{J}$	$\xi_{\mathscr{J}_{min}}$	$\mathscr{G}$	$\xi_{\mathscr{G}_{max}}$	$\dfrac{\mathscr{J} - \mathscr{G}}{\mathscr{G}}$
1	0.5	1.1459	0.881	1.1452	0.880	0.0006
1	1	0.9487	0.596	0.9389	0.580	0.010
−0.5	0.5	0.7006	0.894	0.6988	0.892	0.003
−0.5	1	0.5703	0.694	0.5582	0.678	0.022
−0.5	2	0.3729	0.346	0.3322	0.289	0.123

The values of $\xi_{\mathscr{J}_{min}}$ and $\xi_{\mathscr{G}_{max}}$ for the closest bounds are nearly equal suggesting that the variational solution gives a good approximation to the concentration profile within the slab. The crude bounds (4.4.17) on the integral M are $-0.2274 < M < 0$ for $\kappa = 1$ and $0 < M < 0.2274$ for $\kappa = -0.5$. Thus the information on the effectiveness factor η is rather coarse, e.g., for $\kappa = \Lambda = 1, 0.7115 < \eta < 0.9487$ or for $\kappa = -0.5$, $\Lambda = 1, 0.5582 < \eta < 0.7977$. However if we assume that the concentration profile has been determined with some accuracy, good estimates of M should be possible using the trial functions. Table 4.4.2 gives the bounds on η when M is calculated using the variational approximation.

4.4c. The Variational Estimate of an Upper Bound When Λ is Large

For large values of Λ the profile of concentration falls very steeply within the slab so that c is very small in the central part. A suitable

Table 4.4.2. Estimated bounds on η, $N = 100$.

κ	Λ	$\mathscr{J} + M$	$\mathscr{G} + M$
1	0.5	0.96	0.96
1	1	0.84	0.83
−0.5	0.5	0.87	0.86
−0.5	1	0.66	0.65
−0.5	2	0.41	0.37

trial function, with two disposable parameters, is the even extension into $(-1, 1)$ of

$$u = \begin{cases} 0 & 0 \le \tilde{x} < \beta \\ \chi c_f \{(\tilde{x} - \beta)/(1 - \beta)\}^2 & \beta \le \tilde{x} \le 1 \end{cases} \quad (4.4.32)$$

This function has the desired properties without being unduly complicated, and the values of the parameters are easy to assign by optimization of the bounds. When used in (4.2.8) it gives the upper bound

$$\mathscr{J} = \mathcal{Q}(1 - \beta)^{-1} + \mathscr{T}(1 - \beta) + Y \quad (4.4.33)$$

where

$$\mathcal{Q} = 4\chi^2/3\Lambda^2 \quad (4.4.34)$$

$$\mathscr{T} = \frac{2\chi(1 + \kappa)}{3\kappa} - \frac{2(1 + \kappa)}{\kappa^2} \{\ln(1 + \chi\kappa) - 2 + 2W\} \quad (4.4.35)$$

$$Y = N(1 - \chi)^2/\Lambda^2 \quad (4.4.36)$$

and

$$W = W_+ = \sqrt{1/\chi\kappa} \text{ arc tan } \sqrt{\chi\kappa}, \qquad \kappa > 0$$

$$W = W_- = \sqrt{-1/(\chi\kappa)} \text{ arc tanh } \sqrt{-\chi\kappa}, \qquad -1 < \kappa < 0 \quad (4.4.37)$$

Since $\mathcal{Q}$, $\mathscr{T}$ and Y are functions only of χ, the minimization with respect to β is immediate and gives

$$\beta = 1 - \sqrt{\mathcal{Q}/\mathscr{T}}, \qquad \mathscr{J} = 2\sqrt{\mathcal{Q}\mathscr{T}} + Y \quad (4.4.38)$$

Beyond this point however analytical methods fail and the minimizing value of χ must be found numerically. Table 4.4.3 gives some typical results.

4.4d. The Variational Estimate of a Lower Bound for Large Λ

As in the case of small Λ a suitable trial function for the lower bound is obtained from the derivative of u, namely, the odd extension into $(-1, 1)$ of

$$\mathbf{q} = \begin{cases} 0 & 0 \le \tilde{x} < \beta \\ (4D_e c_f/L)\chi(\tilde{x} - \beta)(1 - \beta)^{-2} \mathbf{i}_x & \beta \le \tilde{x} \le 1 \end{cases} \quad (4.4.39)$$

Table 4.4.3. Bounds for $(\eta - M)$ with $N = 100$.

κ	Λ	$\mathscr{J}$	$\chi_{\mathscr{J}_{min}}$	$\beta_{\mathscr{J}_{min}}$	$\mathscr{G}$	$\chi_{\mathscr{G}_{max}}$	$\beta_{\mathscr{G}_{max}}$	$(\mathscr{J} - \mathscr{G})/\mathscr{G}$
1	5	0.2312	0.95	0.57	0.2040	0.755	0.69	0.133
1	10	0.1100	0.90	0.78	0.0975	0.72	0.85	0.128
1	20	0.0501	0.81	0.89	0.0448	0.65	0.92	0.12
1	50	0.0157	0.62	0.96	0.0143	0.51	0.97	0.10
-0.5	5	0.1612	0.96	0.36	0.1300	0.69	0.60	0.240
-0.5	20	0.0354	0.85	0.84	0.0294	0.63	0.90	0.20
-0.5	50	0.0115	0.70	0.93	0.0099	0.54	0.96	0.16
-0.5	100	0.0044	0.55	0.97	0.0039	0.44	0.98	0.12

For convenience let

$$Z = 2\chi(1 - \beta)^{-2} \tag{4.4.40}$$

then

$$\mathscr{G}(q) = 2(1 - \beta)(1 + \kappa)\kappa^{-1}\{(Z/\Lambda^2) - (Z^2/h_z)$$
$$+ \kappa^{-1}\ln[1 - \kappa Z/\Lambda^2(1 + \kappa)]\} \tag{4.4.41}$$

where

$$h_z = 6\Lambda^2 N(1 + \kappa)\kappa^{-1}(1 - \beta)^{-1}\{3 + N(1 - \beta)\}^{-1} \tag{4.4.42}$$

Again, for positive κ the integral I in (4.2.16) will only exist if

$$Z < \Lambda^2(1 + \kappa)/\kappa \tag{4.4.43}$$

while for negative κ there is no such constraint. As in the section 4.4b the maximizing value of Z satisfies a quadratic

$$Z^2 - \tilde{l}_z Z + \tfrac{1}{2}h_z = 0 \tag{4.4.44}$$

with

$$\tilde{l}_z = \frac{h_z}{2\Lambda^2} + \frac{\Lambda^2(1 + \kappa)}{\kappa} \tag{4.4.45}$$

The appropriate root is given by

$$Z = \tfrac{1}{2}\tilde{l}_z - (\text{sgn } \kappa)\tfrac{1}{2}\sqrt{\tilde{l}_z^2 - 2h_z}. \tag{4.4.46}$$

With this value of Z, $\mathscr{G}$ depends only on the parameter β, the best value of which must be found numerically. Values of the maximizing χ and β and the corresponding lower bound are given in Table 4.4.3. It is noticeable that the relative error is much greater though of course the magnitude of $(\eta - M)$ is smaller. Table 4.4.4 shows what can be said about the estimated bounds on the effectiveness factor by using the best trial function in the integral M.

Table 4.4.4. Estimated bounds on η, $N = 100$

κ	Λ	$\mathscr{J} + M$	$\mathscr{G} + M$
1	5	0.22	0.20
1	10	0.10	0.095
1	20	0.048	0.044
1	50	0.015	0.014
-0.5	5	0.17	0.13
-0.5	20	0.038	0.030
-0.5	50	0.012	0.010
-0.5	100	0.0045	0.0040

4.4.A. Appendix 4.4.1. Integrals

The following integrals may be helpful in the derivation of Eqs. (4.4.21) and (4.4.27).

$$\int \ln (\gamma_1 + \mu_1 \zeta^2)\, d\zeta = \zeta \ln (\gamma_1 + \mu_1 \zeta^2) - 2\zeta$$
$$+ 2(\gamma_1/\mu_1)^{1/2} \tan^{-1}\left((\mu_1/\gamma_1)^{1/2}\, \zeta\right) \quad (4.4.47)$$

and

$$\int \ln (\gamma_1 - \mu_1 \zeta^2)\, d\zeta = \zeta \ln (\gamma_1 - \mu_1 \zeta^2) - 2\zeta$$
$$+ 2(\gamma_1/\mu_1)^{1/2} \tanh^{-1}\left((\mu_1/\gamma_1)^{1/2}\, \zeta\right) \quad (4.4.48)$$

where γ_1 and μ_1 are positive and non-negative constants respectively.

Appendix 4.4.2. Error Bounds on the Variational Solutions

An investigation of the error of the "approximate variational solutions" we have obtained in this section is of some interest. In general, error estimates of this kind are not available, but for certain classes of boundary value problems, error bounds have been derived by Arthurs and Coles (1971) for homogeneous boundary conditions and by Arthurs (1973) for more general boundary conditions. It is the latter work of Arthurs (Arthurs, [5]) that will be applied in this appendix. The derivation is based on the theory of complementary variational principles.

When the complementary upper (4.2.8) and lower (4.2.15) bound variational principles hold, we have shown in (4.2.14), (4.2.23) and (4.2.28) that

$$\mathscr{G}(q) \leq \mathscr{G}_{\max} = \mathscr{J}_{\min} \leq \mathscr{J}(u) \quad (4.4.49)$$

for any admissible trial functions u and q, and we can write

$$\mathscr{J}(u) - \mathscr{G}(q) \geq \mathscr{J}(u) - \mathscr{J}_{\min} \qquad (4.4.50)$$

However, from (4.2.12) and (4.2.13) we find that the difference of $\mathscr{J}$ and its extremum value is in fact the second variation in $\mathscr{J}$

$$\mathscr{J}(u) - \mathscr{G}(q) \geq \mathscr{J}_2(u) \qquad (4.4.51)$$

where

$$Vc_f r_f \mathscr{J}_2 = \int_{\mathscr{V}} \left[D_e (\nabla u_1)^2 + (u_1)^2 r'(c + \theta_m u_1) \right] d^3 \mathbf{x}$$

$$+ \int_{\partial \mathscr{V}} k_c (u_1)^2 \, d^2 \mathbf{x} \qquad (4.4.52)$$

Recall that u_1 is the difference between the trial concentration u and its true value c, and θ_m is a parameter

$$0 < \theta_m < 1.$$

If we apply Gauss' theorem to the ∇u_1 term in (4.4.52) we obtain

$$Vc_f r_f \mathscr{J}_2 = \int_{\mathscr{V}} u_1 \left[-D_e \nabla^2 u_1 + u_1 r'(c + \theta_m u_1) \right] d^3 \mathbf{x}$$

$$+ \int_{\partial \mathscr{V}} u_1 (\mathbf{n} \cdot D_e \nabla u_1 + k_c u_1) \, d^2 \mathbf{x}. \qquad (4.4.53)$$

The continuity of u and its normal derivative across any arbitrary surface within $\mathscr{V}$ as well as a piecewise continuity of second derivatives of u is required for (4.4.53).

To derive a useful formula from this we must consider trial concentrations u which make the boundary terms in (4.4.53) vanish. If we select a trial function u that satisfies the boundary equation (4.2.2)

$$D_e \mathbf{n} \cdot \nabla u = k_c (c_f - u) \quad \text{on } \partial \mathscr{V} \qquad (4.4.54)$$

then as c must also satisfy this boundary condition, so must the difference $u_1 = u - c$ be given by

$$D_e \mathbf{n} \cdot \nabla u_1 = -k_c u_1 \quad \text{on } \partial \mathscr{V}, \qquad (4.4.55)$$

and (4.4.53) becomes

$$Vc_f r_f \mathscr{J}_2 = \int_{\mathscr{V}} u_1 \left[-D_e \nabla^2 u_1 + u_1 r'(c + \theta_m u_1) \right] d^3 x. \qquad (4.4.56)$$

If ω_0^2 denotes the lowest eigenvalue of the positive eigenvalues of the equation

$$\nabla^2 f_p + \omega_p^2 f_p = 0 \quad \text{in } \mathscr{V} \qquad (4.4.57)$$

subject to the boundary condition

$$D_e \boldsymbol{n} \cdot \nabla f_p = - k_c f_p \quad \text{on } \partial \mathscr{V} \tag{4.4.58}$$

and if $\gamma_{\min}$ is a constant such that

$$r'(c + \theta_m u_1) \geq \gamma_{\min} \geq 0 \tag{4.4.59}$$

it follows from (4.4.51) and (4.4.56) that

$$\mathscr{J}(u) - \mathscr{G}(q) \geq (c_f r_f)^{-1} (D_e \omega_0^2 + \gamma_{\min}) \, \|u - c\|_{\mathscr{V}}^2 \tag{4.4.60}$$

where

$$\|u - c\|_{\mathscr{V}} = \left\{ V^{-1} \int_{\mathscr{V}} (u - c)^2 \, d^3 \boldsymbol{x} \right\}^{1/2} . \tag{4.4.61}$$

In terms of the Thiele Modulus,

$$\Lambda^2 = L^2 r_f / 4 D_e c_f, \tag{4.4.62}$$

the mass transfer Biot number

$$N = L k_c / 2 D_e \tag{4.4.63}$$

and the dimensionless quantities

$$\tilde{\omega}_p^2 = \omega_p^2 L^2 / 4 \tag{4.4.64}$$

$$\tilde{\gamma}_{\min} = \gamma_{\min} c_f / r_f \tag{4.4.65}$$

$$\tilde{u} = u / c_f \tag{4.4.66}$$

$$\tilde{c} = c / c_f \tag{4.4.67}$$

we have the final result that

$$\|\tilde{u} - \tilde{c}\|_{\mathscr{V}} \leq \left[\frac{\mathscr{J}(u) - \mathscr{G}(q)}{\tilde{\omega}_0^2 \Lambda^{-2} + \tilde{\gamma}_{\min}} \right]^{1/2} . \tag{4.4.68}$$

For Langmuir-Hinshelwood kinetics (4.4.3)

$$r(c) = \frac{k_r' c}{1 + Kc} \tag{4.4.69}$$

the value of $\tilde{\gamma}_{\min}$ depends on the sign of the dimensionless number $\kappa(= Kc_f)$, we have

$$\tilde{\gamma}_{\min} = (1 + \kappa)^{-1} \qquad 0 < \kappa \tag{4.4.70}$$

and

$$\tilde{\gamma}_{\min} = (1 + \kappa) \qquad 0 > \kappa > -1. \tag{4.4.71}$$

Furthermore the lowest eigenvalue of the equation for a slab

$$\frac{d^2 f_p}{dx^2} + \omega_p^2 f_p = 0 \qquad -L/2 \leq x \leq L/2 \tag{4.4.72}$$

subject to the boundary conditions

$$\frac{df_p}{dx} = \mp \frac{k_c}{D_e} f_p \qquad x = \pm L/2 \qquad (4.4.73)$$

is the smallest positive root of the equation

$$N^{-1}\tilde{\omega}_p = \cot \tilde{\omega}_p \qquad (4.4.74)$$

Trial upper and lower bound solutions were used in the calculation of the estimated bounds on the effectiveness factor η in Tables 4.4.2 and 4.4.4. Though rigorous bounds (4.4.15) on $\eta - M$ were provided by $\mathscr{G}$ and $\mathscr{J}$, M, as given by

$$M = \frac{1}{Lc_f r_f} \int_{-L/2}^{L/2} \left[cr(c) - 2 \int_0^c r(\zeta)\,d\zeta \right] dx, \qquad (4.4.75)$$

could only be estimated by substituting upper and lower bound variational trial functions in the place of c. Unfortunately, it is difficult to use a bound on the error as measured by $\|\tilde{u} - \tilde{c}\|_{\mathscr{V}}$ of (4.4.61) to bound the error in the estimated M. The results of Table 4.4.5 do however indicate for small Λ the usefulness of the variational solution form.

Table 4.4.5. Upper Bounds on the Error of the Variational Solution for $N = 100$

κ	Λ	Upper Bound on $\|\tilde{u} - \tilde{c}\|_{\mathscr{V}}$
1	0.5	0.009
1	1	0.058
−0.5	0.5	0.014
−0.5	1	0.065
−0.5	2	0.192

4.5. Effects of Pellet Shape on the Effectiveness Factor

In this section we shall show that of all catalyst pellets of fixed volume in which an isothermal first-order reaction is taking place, the spherical particle has the least effectiveness factor. This is not a surprising result for the sphere having the least surface area per unit volume of all shapes has, in a loose manner of speaking, the most inaccessible interior. This inaccessibility militates against obtaining a fast overall reaction rate for this particular form of kinetics. However the result, though not unexpected, is not trivial for there are other circumstances in which the inaccessibility would actually enhance the rate. The concepts introduced here differ somewhat from the usual idea of a "shape factor" in so far as comparisons are being made between shapes having the same volume, rather than shapes having *one* dimension the same. Indeed the infinite cylinder and slab must be excluded since they do not have a finite volume. The discussion follows arguments advanced by Luss and Amundson (1967).

For simplicity these authors assume that external mass transfer can be ignored ($k_c \to \infty$) and that Dirichlet conditions are applied on the external boundary $\partial \mathcal{V}$ of the catalyst pellet

$$c = c_f \quad \text{on } \partial \mathcal{V} \tag{4.5.1}$$

Within the pellet the reaction is of the first order and from (4.2.1) we have

$$\nabla^2 c = \phi_c^2 c \tag{4.5.2}$$
$$\phi_c^2 = k/D$$

where k is the reaction rate constant per unit volume, and D is the effective diffusion coefficient. The effectiveness factor η from (4.2.4) is

$$\eta = \int_{\mathcal{V}} c \, d^3 x / V c_f \tag{4.5.3}$$

Aris (1957) gave a formula for the effectiveness factor of a finite cylinder of height h and radius r, and others have considered various shapes such as finite and hollow cylinders (Gunn, 1967) and parallelepipeds (Luss and Amundson, 1967); see also Aris (1974) and Kohn and Strieder (1973). Before proceeding with a proof we will examine the effectiveness factor for a finite cylinder for various diameter to height ratios $\gamma^3 = 2r/h$ but with a fixed volume

$$\eta_c(\text{cylinder}) = 1 - \frac{16\phi_c^2 r^2}{\pi^2} \sum_{p=0}^{\infty} \frac{1}{(2p+1)^2 \alpha_p^2} \left[\frac{1}{2} - \frac{I_1(\alpha_p)}{\alpha_p I_0(\alpha_p)} \right]$$

where

$$\alpha_p^2 = \phi_c^2 r^2 + (2p+1)^2 \pi^2 \gamma^6 / 4, \tag{4.5.4}$$

and $I_0(\alpha_p)$, $I_1(\alpha_p)$ are modified Bessel functions. In Fig. 4.5.1 the effectiveness factor η_c is plotted versus

$$\phi = R\phi_c \tag{4.5.5}$$

for different values of the ratio $\gamma = (2r/h)^{1/3}$ where r the radius of the cylinder is related to a sphere of radius R with the same volume by $r = \gamma R (2/3)^{1/3}$. At a given value of the parameter ϕ the corresponding η_c's are all for the same volume of particle. The sphere is the lowest curve, and $\gamma = 1$ corresponds to the cylinder with least surface for a

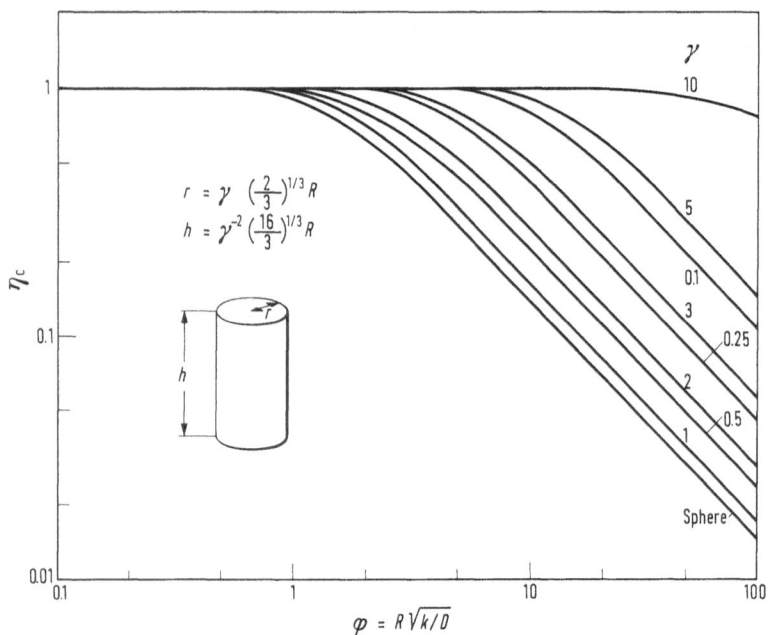

Fig. 4.5.1. Plot of effectiveness factor η_c for a cylinder vs. ϕ where R is the radius of a sphere having the same volume. (Note that this ϕ is $R\phi_c$).

fixed volume. The effectiveness factors tend to increase as the cylinder gets flatter or longer. There is clearly no unique shape to maximize η for a given volume, for a large and very thin slab will give η as close to 1 as we please. However there is a shape that minimizes η, namely the sphere, as we shall now demonstrate. The proof is accomplished by a mathematical process known as symmetrization invented by Jacob Steiner in 1836 and generalized and elaborated by many (Polya and Szego, 1951).

The variational upper bound (4.2.8) written in the limit of perfect external mass transfer ($k_c \to \infty$), subject to the condition (4.2.9), provides an upper bound (4.2.14) on the effectiveness factor (4.2.28) for any simply connected region

$$Vc_f^2\phi_c^2\eta \leq \int_{\mathscr{V}} \{(\nabla u)^2 + \phi_c^2 u^2\} \, d^3x \qquad (4.5.6)$$

Note that the equality holds if u is the solution of (4.5.2). We attempt to construct a function $\bar{c}$ in the sphere $\mathscr{M}$ of volume V where $\bar{c} = c_f$ on the surface $\partial\mathscr{M}$ of the sphere, and

$$\int_{\mathcal{M}} \{(\nabla \bar{c})^2 + \phi_c^2 \bar{c}^2\} \, d^3 x < \int_{\mathcal{V}} \{(\nabla c)^2 + \phi_c^2 c^2\} \, d^3 x, \qquad (4.5.7)$$

where c is the solution of (4.5.1) and (4.5.2). Then the equality (4.2.28) relates the right-hand side of (4.5.7) to the effectiveness factor $\eta_{\mathcal{V}}$ of the arbitrary volume $\mathcal{V}$, and the inequality (4.5.6) tells us that the left-hand side of (4.5.7) is an upper bound on the effectiveness factor $\eta_{\mathcal{M}}$ for a sphere $\mathcal{M}$ of volume V

$$\eta_{\mathcal{M}} < \eta_{\mathcal{V}} \qquad (4.5.8)$$

For a given volume η has a minimum value for the spherical shape, as is illustrated in Fig. 4.5.1.

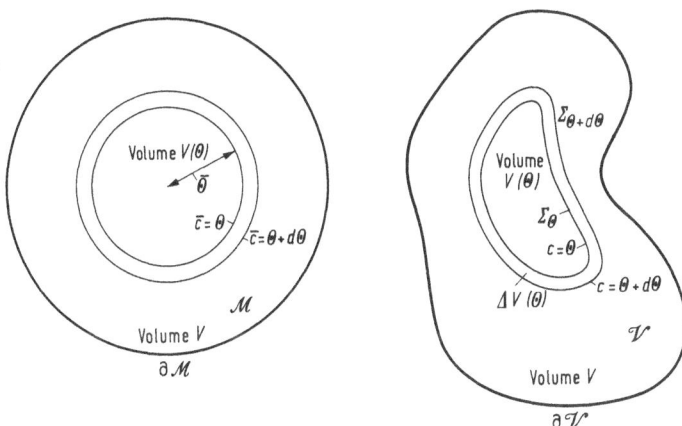

Fig. 4.5.2. The symmetrization process.

A function which satisfies condition (4.5.7) may be constructed by spherical symmetrization. In the arbitrary body $\mathcal{V}$ we can construct level surfaces of concentration. These surfaces Σ_θ are $c = \theta, 0 \le \theta \le c_f$. The volume inside Σ_θ is $V(\theta)$, and the domain between two neighboring level surfaces $c = \theta$ and $c = \theta + d\theta$ will be called $\Delta \mathcal{V}(\theta)$.

The function $\bar{c}(\bar{\theta})$ will be chosen such that the volume enclosed by any level surface in the sphere and the body $\mathcal{V}$ will be the same. The volume of the sphere and the body are defined by the level surface $c = \bar{c} = c_f$ to be the same. The function $\bar{c}$ will be constructed so that

$$\bar{c}(\bar{\theta}) = \theta \qquad (4.5.9)$$

where $\bar{\theta}$ is given by

$$\tfrac{4}{3}\pi \bar{\theta}^3 = V(\theta). \qquad (4.5.10)$$

Define a function

$$G_\theta = (\nabla c \cdot \boldsymbol{n}) = \frac{d\theta}{dn} = |\nabla c| \qquad (4.5.11)$$

since ∇c and $\boldsymbol{n}$ are both normal to the level surfaces of c. Now, recalling that $d^2\boldsymbol{x}$ denotes the surface element of the surface named at the integral sign,

$$\int_{\Delta \mathscr{V}(\theta)} (\nabla c)^2 \, d^3\boldsymbol{x} = \int_{\Sigma_\theta} (\nabla c)^2 \, d^2\boldsymbol{x} \, dn = \int_{\Sigma_\theta} G_\theta^2 \, d^2\boldsymbol{x} \, dn$$

$$= \int_{\Sigma_\theta} G_\theta \frac{d\theta}{dn} d^2\boldsymbol{x} \, dn = d\theta \int_{\Sigma_\theta} G_\theta \, d^2\boldsymbol{x} = \bar{P}(\theta) \, d\theta, \quad (4.5.12)$$

where

$$\bar{P}(\theta) = \int_{\Sigma_\theta} G_\theta \, d^2\boldsymbol{x} \qquad (4.5.13)$$

The difference in the volumes of neighboring domains is

$$V(\theta + d\theta) - V(\theta) = V'(\theta) \, d\theta$$

$$= \int_{\Delta \mathscr{V}(\theta)} d^3\boldsymbol{x} = \int_{\Sigma_\theta} d^2\boldsymbol{x} \, dn = \int_{\Sigma_\theta} d^2\boldsymbol{x} \frac{d\theta}{G_\theta} \qquad (4.5.14)$$

or

$$V'(\theta) = \int_{\Sigma_\theta} G_\theta^{-1} \, d^2\boldsymbol{x} \qquad (4.5.15)$$

From the Schwartz inequality

$$\left(\int_{\Sigma_\theta} G_\theta \, d^2\boldsymbol{x} \right) \left(\int_{\Sigma_\theta} G_\theta^{-1} \, d^2\boldsymbol{x} \right) \geq \left(\int_{\Sigma_\theta} d^2\boldsymbol{x} \right)^2 \qquad (4.5.16)$$

$$= \{\text{surface of } \Sigma_\theta\}^2$$

The isoperimetric inequality for a simply connected body of volume V states (Polya and Szego, 1951) that of all possible shapes the sphere has the smallest surface area. Hence

$$S_{\partial \mathscr{V}} \geq 4\pi \left(\frac{3V}{4\pi} \right)^{2/3} = H_s V^{2/3} \qquad (4.5.17)$$

where

$$H_s = (4\pi)^{1/3} \, 3^{2/3} \, .$$

Obviously the equality (4.5.17) holds for a sphere. From equations

(4.5.13), (4.5.15), (4.5.16) and (4.5.17)

$$\bar{P}(\theta)\,V'(\theta) \geq \Sigma_\theta^2 \geq H_s^2\,\{V(\theta)\}^{4/3} \tag{4.5.18}$$

or

$$\bar{P}(\theta) \geq H_s^2 V(\theta)^{4/3}/V'(\theta) \tag{4.5.19}$$

Note that G_θ is a constant for the sphere, the equalities hold in (4.5.16), (4.5.17), and consequently in (4.5.19). Hence

$$\int_{\mathscr{V}} (\nabla c)^2\,d^3x = \int_0^{c_f} \bar{P}(\theta)\,d\theta \geq \int_0^{c_f} H_s^2\,\frac{V(\theta)^{4/3}}{V'(\theta)}\,d\theta \tag{4.5.20}$$

Now the spherical symmetrization is such that in the sphere and in the arbitrary volume $\mathscr{V}$ both $V(\theta)$ and $V'(\theta)$ are the same. Thus it follows from (4.5.20) that

$$\int_{\mathscr{V}} (\nabla c)^2\,d^3x > \int_{\mathscr{M}} (\nabla\bar{c})^2\,d^3x \tag{4.5.21}$$

From the definition of $\bar{c}$

$$\int_{\mathscr{V}} c^2\,d^3x = \int_{\mathscr{M}} \bar{c}^2\,d^3x \tag{4.5.22}$$

and from (4.5.21) and (4.5.22) we obtain

$$\int_{\mathscr{M}} \{(\nabla\bar{c})^2 + \phi_c^2\bar{c}^2\}\,d^3x < \int_{\mathscr{V}} \{(\nabla c)^2 + \phi_c^2 c^2\}\,d^3x.$$

This is the inequality (4.5.7), from which the minimal property (4.5.8) is derived.

Luss and Amundson (1967) went on to show that the same reasoning applies to reversible, isothermal, first-order kinetics.

4.6. Variational Principles for Heterogeneous Catalysis: the Discrete Model

In conventional catalysts the active sites are sufficiently distributed to justify the assumption of homogeneity, but in the molecular sieve type catalyst the reaction is localized and its discrete nature should be recognized. We consider a catalyst pellet (Fig. 4.6.1) of volume V, and let $\hat{\mathscr{V}}$ denote an inert porous region within the external surface $\partial\mathscr{V}$ through which reactants and products must diffuse. While the porous region $\hat{\mathscr{V}}$ is inert, the reaction takes place on or in a number of finer particles distributed throughout the pellet. $\partial\hat{\mathscr{V}}$ denotes the union of

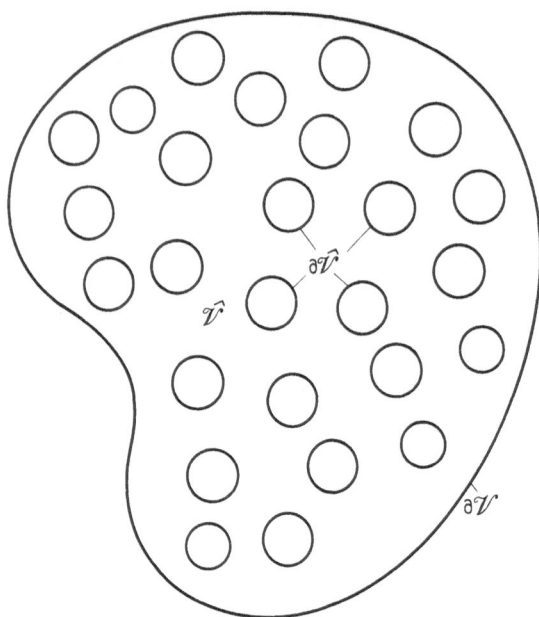

Fig. 4.6.1. A catalyst pellet with discrete reactive surfaces.

all the internal reaction surfaces. For simplicity we will assume none of the sieves protrudes through the outer surface $\partial \mathcal{V}$. Then if

$$\hat{r}(c) = \text{rate of reaction per unit area of } \partial \hat{\mathcal{V}}$$

$$\hat{S} = \text{total internal catalytic area}$$

we have

$$\nabla \cdot (D_e \nabla c) = 0 \qquad \text{in } \mathcal{V} \qquad (4.6.1)$$

$$D_e \boldsymbol{n} \cdot \nabla c = k_c(c_f - c) \quad \text{on } \partial \mathcal{V} \qquad (4.6.2)$$

$$D_e \boldsymbol{n} \cdot \nabla c = \hat{r}(c) \qquad \text{on } \partial \hat{\mathcal{V}} \qquad (4.6.3)$$

where $\boldsymbol{n}$ is the outward unit normal vector (4.6.2) pointing away from the catalyst surface on $\partial \mathcal{V}$, and (4.6.3) pointing into the inert region $\hat{\mathcal{V}}$ on the molecular sieve surface $\partial \hat{\mathcal{V}}$. The remaining symbols are defined in section 4.2.

The effectiveness factor $\hat{\eta}$, defined as the ratio of the rate of reaction within the catalyst to the rate of reaction within the catalyst with no diffusion restriction, can be expressed in two forms

$$\hat{\eta} = \int_{\partial \hat{\mathcal{V}}} \hat{r}(c) \, d^2 \boldsymbol{x} / \hat{S} \hat{r}_f \qquad (4.6.4)$$

or

$$\hat{\eta} = \int_{\partial \hat{\mathscr{V}}} D_e \boldsymbol{n} \cdot \nabla c \ d^2 \boldsymbol{x} / \hat{S} \hat{r}_f \qquad (4.6.5)$$

where

$$\hat{r}_f = \hat{r}(c_f).$$

In addition we define the following dimensionless quantities: a characteristic length $L/2$ is the total pellet volume to external surface area ratio, the Thiele modulus is defined by

$$\hat{\Lambda}^2 = L\hat{r}_f / 2D_e c_f \qquad (4.6.6)$$

a dimensionless surface area for the catalytically active surface is given by

$$\hat{\sigma} = 4\hat{S}/L^2 \qquad (4.6.7)$$

and as in the homogeneous case we can use

$$N = Lk_c / 2D_e \qquad (4.2.7 \text{ bis})$$

for the external mass transfer.

4.6a. Upper Bound

Upper and lower bound principles have been recently formulated by Rester and Aris (1972). If u is some function which is twice differentiable in $\hat{\mathscr{V}}$ and satisfies

$$|k_c(c_f - u)| < \infty \quad \text{on } \partial \hat{\mathscr{V}} \qquad (4.6.8)$$

it may be used as a trial function for the concentration in the variational functional

$$\hat{S}c_f \hat{r}_f \hat{\mathscr{J}}(u) = \int_{\hat{\mathscr{V}}} D_e \nabla u \cdot \nabla u \ d^3 \boldsymbol{x} + 2 \int_{\partial \hat{\mathscr{V}}} \hat{F}(u) \ d^2 \boldsymbol{x}$$

$$+ \int_{\partial \hat{\mathscr{V}}} k_c(c_f - u)^2 \ d^2 \boldsymbol{x} \qquad (4.6.9)$$

where

$$\hat{F}(u) = \int_0^u \hat{r}(\zeta) \ d\zeta \qquad (4.6.10)$$

We consider the variations u_1 around u_0, and write the term in (4.6.9) which is linear in u_1 as

$$\hat{S}c_f \hat{r}_f \hat{\mathscr{J}}_1 = 2 \int_{\hat{\mathscr{V}}} D_e \nabla u_1 \cdot \nabla u_0 \ d^3 \boldsymbol{x} + 2 \int_{\partial \hat{\mathscr{V}}} \hat{r}(u_0) \ u_1 \ d^2 \boldsymbol{x}$$

$$- 2 \int_{\partial \mathscr{V}} u_1 k_c (c_f - u_0) \, d^2 \boldsymbol{x}, \qquad (4.6.11)$$

$$= - 2 \int_{\mathscr{V}} u_1 \nabla \cdot (D_e \nabla u_0) \, d^3 \boldsymbol{x}$$

$$+ 2 \int_{\partial \mathscr{V}} u_1 \{ D_e \boldsymbol{n} \cdot \nabla u_0 - k_c (c_f - u_0) \} \, d^2 \boldsymbol{x}$$

$$+ 2 \int_{\partial \mathscr{V}} u_1 \{ - D_e \boldsymbol{n} \cdot \nabla u_0 + \hat{r}(u_0) \} \, d^2 \boldsymbol{x} \qquad (4.6.12)$$

The second expression follows from the application of Gauss' theorem, and several elementary manipulations. At an extremum the first variation $\hat{\mathscr{J}}_1$ must vanish for any u_1 and this will be the case if the u_0 in (4.6.12) is the solution of (4.6.1), (4.6.2), and (4.6.3), i.e., if $u_0 = c$. After some rearrangement the variational functional (4.6.9) can be rewritten

$$\hat{\mathscr{J}}(c + u_1) = \hat{\mathscr{J}}(c) + \hat{\mathscr{J}}_2 \qquad (4.6.13)$$

where

$$\hat{S} c_f \hat{r}_f \hat{\mathscr{J}}_2 = \int_{\mathscr{V}} D_e (\nabla u_1)^2 \, d^3 \boldsymbol{x} + \int_{\partial \mathscr{V}} u_1^2 \hat{r}'(c + \theta_m u_1) \, d^2 \boldsymbol{x}$$

$$+ \int_{\partial \mathscr{V}} k_c u_1^2 \, d^2 \boldsymbol{x} \qquad (4.6.14)$$

and $0 < \theta_m < 1$. If the reaction rate is monotonic increasing $\hat{r}' = d\hat{r}/d\zeta$ is always positive. So also is $\hat{\mathscr{J}}_2$ and hence

$$\hat{\mathscr{J}}(u) \geq \hat{\mathscr{J}}(c) \qquad (4.6.15)$$

4.6b. Lower Bound

The lower bound variational principle is written in terms of a trial flux vector $\boldsymbol{q}$, which is continuously differentiable and satisfies the constraint

$$\nabla \cdot \boldsymbol{q} = 0 \qquad \text{in } \mathscr{V} \qquad (4.6.16)$$

The variational principle is

$$\hat{S} c_f \hat{r}_f \hat{\mathscr{G}}(\boldsymbol{q}) = - \int_{\mathscr{V}} \left\{ \frac{\boldsymbol{q} \cdot \boldsymbol{q}}{D_e} + 2\omega \, \nabla \cdot \boldsymbol{q} \right\} d^3 \boldsymbol{x}$$

$$- 2 \int_{\partial \mathscr{V}} \hat{I} (\boldsymbol{n} \cdot \boldsymbol{q}) \, d^2 \boldsymbol{x}$$

$$+ \int_{\partial \mathscr{V}} \left\{ 2 c_f \boldsymbol{n} \cdot \boldsymbol{q} - \frac{(\boldsymbol{n} \cdot \boldsymbol{q})^2}{k_c} \right\} d^2 \boldsymbol{x} \qquad (4.6.17)$$

with

$$\hat{I}(\boldsymbol{n} \cdot \boldsymbol{q}) = \int_0^{\boldsymbol{n} \cdot \boldsymbol{q}} \hat{r}^{-1}(\zeta) \, d\zeta \qquad (4.6.18)$$

where $\hat{r}^{-1}$ is the inverse of the reaction rate function in (4.6.3), and $\omega(\boldsymbol{x})$ is an undetermined Lagrangian multiplier to account for the constraint (4.6.16) on $\boldsymbol{q}$.

The trial flux vector $\boldsymbol{q}_0 + \boldsymbol{q}_1$ is introduced into (4.6.17) and the terms of the first order in the variation $\boldsymbol{q}_1$ are collected:

$$\begin{aligned}
\hat{S}c_f \hat{r}_f \hat{\mathscr{G}}_1 = & -2 \int_{\hat{\mathscr{V}}} \left\{ \frac{\boldsymbol{q}_1 \cdot \boldsymbol{q}_0}{D_e} + \omega \, \nabla \cdot \boldsymbol{q}_1 \right\} d^3 \boldsymbol{x} \\
& -2 \int_{\partial \hat{\mathscr{V}}} (\boldsymbol{n} \cdot \boldsymbol{q}_1) \hat{r}^{-1}(\boldsymbol{n} \cdot \boldsymbol{q}_0) \, d^2 \boldsymbol{x} \\
& +2 \int_{\partial \hat{\mathscr{V}}} \boldsymbol{n} \cdot \boldsymbol{q}_1 \left\{ c_f - \frac{(\boldsymbol{n} \cdot \boldsymbol{q}_0)}{k_c} \right\} d^2 \boldsymbol{x} \qquad (4.6.19) \\
= & -2 \int_{\hat{\mathscr{V}}} \boldsymbol{q}_1 \cdot \left\{ \frac{\boldsymbol{q}_0}{D_e} - \nabla \omega \right\} d^3 \boldsymbol{x} \\
& + \int_{\partial \hat{\mathscr{V}}} (\boldsymbol{n} \cdot \boldsymbol{q}_1) \left\{ -2 \hat{r}^{-1}(\boldsymbol{n} \cdot \boldsymbol{q}_0) + 2\omega \right\} d^2 \boldsymbol{x} \\
& + \int_{\partial \mathscr{V}} (\boldsymbol{n} \cdot \boldsymbol{q}_1) \left\{ 2c_f - 2\omega - 2 \frac{(\boldsymbol{n} \cdot \boldsymbol{q}_0)}{k_c} \right\} d^2 \boldsymbol{x} \quad (4.6.20)
\end{aligned}$$

where the second expression results from the application of Gauss' theorem. The Euler-Lagrange equations are thus

$$\boldsymbol{q}_0 - D_e \, \nabla \omega = 0 \quad \text{in } \hat{\mathscr{V}} \qquad (4.6.21)$$

$$\hat{r}^{-1}(\boldsymbol{n} \cdot \boldsymbol{q}_0) - \omega = 0 \quad \text{on } \partial \hat{\mathscr{V}} \qquad (4.6.22)$$

and

$$k_c \{c_f - \omega\} - \boldsymbol{n} \cdot \boldsymbol{q}_0 = 0 \quad \text{on } \partial \mathscr{V} \qquad (4.6.23)$$

From (4.6.21) the vector $\boldsymbol{q}_0$ is D_e times the gradient of a scalar ω, and with the constraint (4.6.16)

$$\nabla \cdot \boldsymbol{q}_0 = \nabla \cdot (D_e \, \nabla \omega) = 0 \quad \text{in } \hat{\mathscr{V}}; \qquad (4.6.24)$$

furthermore as $\hat{r}^{-1}$ is the inverse of $\hat{r}$ we have from (4.6.22)

$$\boldsymbol{n} \cdot \boldsymbol{q}_0 = \hat{r}(\omega) \quad \text{on } \partial \hat{\mathscr{V}} \qquad (4.6.25)$$

The Euler-Lagrange equations (4.6.23), (4.6.24), and (4.6.25) are of the same form as those of the discrete model catalyst pellet, (4.6.1), (4.6.2), and (4.6.3). Thus identification of $\omega(x)$ with the reactant concentration c makes the first order terms $\hat{\mathcal{G}}_1$ vanish.

The variational functional $\hat{\mathcal{G}}$ can now be written

$$\hat{\mathcal{G}}(D_e\,\nabla c + \boldsymbol{q}_1) = \hat{\mathcal{G}}(D_e\,\nabla c) + \hat{\mathcal{G}}_2, \tag{4.6.26}$$

where

$$\hat{S}c_f\hat{r}_f\hat{\mathcal{G}}_2 = -\int_{\hat{\mathscr{V}}} \frac{\boldsymbol{q}_1\cdot\boldsymbol{q}_1}{D_e}\,d^3x - \int_{\partial\hat{\mathscr{V}}} (\boldsymbol{n}\cdot\boldsymbol{q}_1)^2\,(\hat{r}^{-1})'\,(\boldsymbol{n}\cdot D_e\,\nabla c + \theta_m\boldsymbol{n}\cdot\boldsymbol{q}_1)\,d^2x$$

$$-\int_{\partial\hat{\mathscr{V}}} \frac{(\boldsymbol{n}\cdot\boldsymbol{q}_1)^2}{k_c}\,d^2x, \tag{4.6.27}$$

and $0 < \theta_m < 1$. $\hat{\mathcal{G}}_2$ is clearly negative for any monotonic increasing $\hat{r}$, since

$$\frac{d\hat{r}^{-1}(\zeta)}{d\zeta} = 1 \left/ \left(\frac{d\hat{r}(v)}{dv} \right) \right. > 0. \tag{4.6.28}$$

The lower bound is therefore

$$\hat{\mathcal{G}}(\boldsymbol{q}) \le \hat{\mathcal{G}}(D_e\,\nabla c). \tag{4.6.29}$$

To equate maximum $\hat{\mathcal{G}}$ to minimum $\hat{\mathcal{J}}$ we first note that $\hat{F}$ from (4.6.10) and $\hat{I}$ from (4.6.18) are related by

$$\hat{I}\{\hat{r}(c)\} = \int_0^{\hat{r}(c)} \hat{r}^{-1}(\zeta)\,d\zeta = c\hat{r}(c) - \hat{F}(c). \tag{4.6.30}$$

The maximum value of $\hat{\mathcal{G}}$ for $\boldsymbol{q} = D_e\,\nabla c$ can be written, with the aid of (4.6.1), (4.6.3), and (4.6.30), in the form

$$\hat{S}c_f\hat{r}_f\hat{\mathcal{G}}_{\max} = -\int_{\hat{\mathscr{V}}} D_e(\nabla c)^2\,d^3x - 2\int_{\partial\hat{\mathscr{V}}} \{c\hat{r}(c) - \hat{F}(c)\}\,d^2x$$

$$+ \int_{\partial\hat{\mathscr{V}}} D_e\boldsymbol{n}\cdot\nabla c\,\{2c_f - (D_e/k_c)(\boldsymbol{n}\cdot\nabla c)\}\,d^2x. \tag{4.6.31}$$

We now apply Gauss' theorem and the differential equation (4.6.1) with boundary conditions (4.6.2) and (4.2.3) to the first integral in (4.6.31) giving

$$\hat{S}c_f\hat{r}_f\hat{\mathcal{G}}_{\max} = \int_{\partial\hat{\mathscr{V}}} \{2\hat{F}(c) - c\hat{r}(c)\}\,d^2x + c_f\int_{\partial\hat{\mathscr{V}}} D_e(\boldsymbol{n}\cdot\nabla c)\,d^2x \tag{4.6.32}$$

The boundary condition (4.6.3) is now substituted into the $\hat{r}$ term in the integral over $\partial\hat{\mathscr{V}}$ and Gauss' theorem is again applied, to give

$$\hat{S} c_f \hat{r}_f \mathscr{G}_{max} = 2 \int_{\partial \mathscr{V}} \hat{F}(c) \, d^2 x + \int_{\mathscr{V}} \nabla \cdot (c D_e \, \nabla c) \, d^3 x$$

$$+ \int_{\partial \mathscr{V}} D_e (\boldsymbol{n} \cdot \nabla c)(c_f - c) \, d^2 x \qquad (4.6.33)$$

Equation (4.6.2) is substituted into the integral over $\partial \mathscr{V}$ to eliminate $D_e (\boldsymbol{n} \cdot \nabla c)$, and (4.6.1) is used to obtain

$$\hat{S} c_f \hat{r}_f \mathscr{G}_{max} = \int_{\mathscr{V}} D_e (\nabla c)^2 \, d^3 x + 2 \int_{\partial \mathscr{V}} \hat{F}(c) \, d^2 x$$

$$+ \int_{\partial \mathscr{V}} k_c (c_f - c)^2 \, d^2 x \qquad (4.6.34)$$

From the definition of the effectiveness factor, (4.6.5) and (4.6.9), (4.6.32), (4.6.34) we have

$$\mathscr{G}_{max} = \hat{\mathscr{J}}_{min} = \hat{\eta} - \frac{1}{c_f \hat{r}_f \hat{S}} \int_{\partial \mathscr{V}} \hat{R}(c) \, d^2 x \qquad (4.6.35)$$

where

$$\hat{R}(c) = c \hat{r}(c) - 2 \hat{F}(c) \qquad (4.6.36)$$

$\hat{R}(c)$ vanishes if $\hat{r}(c)$ is linear, but as in the homogeneous case it can sometimes be bounded.

4.7. Application to Linear Kinetics in a Slab: Discrete Model

To illustrate the application of the variational principle we shall calculate an upper bound on $\hat{\eta}$ for the simplest diffusion reaction problem, a first order, isothermal reaction in a flat slab of thickness L. We let x be the distance from the center plane to the surface, and assume that the effective diffusion coefficient D_e is constant in $\mathscr{V}$. Within the inert volume of the catalyst pellet, internal diffusion is determined by

$$\nabla \cdot (D_e \, \nabla c) = 0 \qquad \text{in } \mathscr{V}, \qquad (4.7.1)$$

external mass transfer follows

$$D_e \boldsymbol{n} \cdot \nabla c = k_c (c_f - c) \qquad \text{on } \partial \mathscr{V} \qquad (4.7.2)$$

while the actual reaction takes place on the surface of the molecular sieve crystals

$$D_e \boldsymbol{n} \cdot \nabla c = \hat{k} c \qquad \text{on } \partial \hat{\mathscr{V}} \qquad (4.7.3)$$

where $\hat{k}$ is the first order rate constant, the unit normal $\boldsymbol{n}$ points outward from $\mathscr{V}$ on $\partial\mathscr{V}$ and into the inert volume $\hat{\mathscr{V}}$ on $\partial\hat{\mathscr{V}}$.

The upper bound on $\hat{\eta}$ may be written in the form

$$\hat{S}c_f\hat{r}_f\hat{\eta} < \int_{\mathscr{V}} D_e(\nabla u)^2\, d^3\boldsymbol{x} + \int_{\partial\hat{\mathscr{V}}} \hat{k}u^2\, d^2\boldsymbol{x}$$

$$+ \int_{\partial\mathscr{V}} k_c(c_f - u)^2\, d^2\boldsymbol{x} \tag{4.7.4}$$

We will assume that the distribution of the molecular sieves is uniform, and neither the volume fraction $\hat{\Phi} = \hat{V}/V$ nor the active surface per unit *total* volume $\hat{s} = \hat{S}/V$ will depend on position. The trial concentration which we will assume depends only on the distance coordinate across the slab

$$u = u(x) \qquad -L/2 \leq x \leq L/2, \tag{4.7.5}$$

and, with this the upper bound (4.7.4) becomes

$$\hat{k}c_f^2 L\hat{s}\hat{\eta} < \int_{-L/2}^{L/2} \left\{ \hat{\Phi}D_e\left(\frac{du}{dx}\right)^2 + \hat{k}\hat{s}u^2 \right\} dx$$

$$+ k_c\{c_f - u(L/2)\}^2 + k_c\{c_f - u(-L/2)\}^2 \tag{4.7.6}$$

The right-hand side of inequality (4.7.6) is a minimum for the trial $u(x)$ that satisfies the Euler-Lagrange equation

$$\frac{d}{dx}\left(\hat{\Phi}D_e\frac{du}{dx} \right) - \hat{s}\hat{k}u = 0 \tag{4.7.7}$$

with the boundary conditions

$$\hat{\Phi}D_e\frac{du}{dx} = \pm\, k_c(c_f - u) \quad \text{at} \quad x = \pm\frac{L}{2} \tag{4.7.8}$$

at the two faces. Further in terms of the trial u, the solution of equations (4.7.7) and (4.7.8), the inequality (4.7.6) can be written

$$\hat{\eta} < \frac{1}{\hat{s}\hat{k}c_f L} \int_{-L/2}^{L/2} \frac{d}{dx}\left(\hat{\Phi}D_e\frac{du}{dx} \right) dx \tag{4.7.9}$$

or from (4.7.7)

$$\hat{\eta} < \frac{1}{\hat{s}\hat{k}c_f L} \int_{-L/2}^{L/2} (\hat{s}\hat{k}u)\, dx \tag{4.7.10}$$

The right-hand side of (4.7.10) is the homogeneous effectiveness factor, for an isothermal, first-order reaction and upon solution of (4.7.7) and

(4.7.8) we have the familiar (4.3.9) result

$$\frac{1}{\hat{\eta}} > \frac{\Lambda_{\hat{\Phi}}^2}{N_{\hat{\Phi}}} + \frac{\Lambda_{\hat{\Phi}}}{\tanh \Lambda_{\hat{\Phi}}} \qquad (4.7.11)$$

where similar to (4.2.6) and (4.2.7) we have defined

$$\Lambda_{\hat{\Phi}}^2 = L^2 \hat{k}\hat{s}/4\hat{\Phi}D_e \qquad (4.7.12)$$

and

$$N_{\hat{\Phi}} = Lk_c/2\hat{\Phi}D_e.$$

The presence of the volume fraction $\hat{\Phi}$ in the upper bound (4.7.11) occurs simply because of the blocking effect of the solid reactive spheres. The result (4.7.11) can be realized in practice by an impenetrable solid slab, through which thin cylinders are drilled to run directly across the slab. The reactant can diffuse (with diffusion coefficient D_e) within the cylinder and undergo a first-order reaction on its internal surfaces. The slab we have just described is statistically homogeneous, then in the spirit of section 2.5 inequality (4.7.11) is the best possible upper bound on $\hat{\eta}$ for a statistically homogeneous slab, and any improvement in (4.7.11) will require additional statistical information about the spatial distribution of the reactive surfaces. Furthermore for a given volume fraction $\hat{\Phi}$ we obtain the result, that a best distribution of reactive surface is in fact the one described directly above for indeed in terms of $\hat{\eta}$ no homogeneous system can exceed it.

Some further results are also given by Rester and Aris (1972).

4.8. Analysis of Experimental Data

For case of a first-order isothermal reaction, given a set of experimental data, a region can be constructed within which all other accurate data points must lie. The procedure will be demonstrated with the discrete model, similar results can be obtained for the homogeneous model. The inequalities presented are those of Rester and Aris (1972). Using the same method Prager (1969) has developed inequalities on the effective magnetic permeability in a random two-phase material.

We recall the dimensionless numbers $\hat{\Lambda}^2$, $\hat{\sigma}$, and N introduced in (4.6.6) and (4.6.7), and also the equations (4.7.1) through (4.7.3) for diffusion and first-order reaction on the catalytically active surfaces distributed throughout the pellet. For first order irreversible surface reaction on the catalytically active surface, the dimensionless numbers become

$$\hat{\Lambda}^2 = L\hat{k}/2D_e, \tag{4.8.1}$$

$$\hat{\sigma} = 4\hat{S}/L^2, \tag{4.6.7 bis}$$

and

$$N = Lk_c/2D_e \tag{4.2.7 bis}$$

where $L/2$ is the total pellet volume to external surface ratio. If we define in addition a dimensionless trial concentration $\tilde{u}$

$$\tilde{u} = u/c_f, \tag{4.8.2}$$

a dimensionless trial flux

$$\tilde{q} = qL/(2D_e c_f) \tag{4.8.3}$$

the length

$$\tilde{x} = 2x/L \tag{4.8.4}$$

and a dimensionless element of internal surface area

$$d^2\tilde{x} = 4d^2x/L^2 \tag{4.8.5}$$

the variational bounds (4.6.9) and (4.6.17) written for a first-order surface reaction in terms of these dimensionless variables become

$$\int_{\mathscr{V}} \frac{1}{\hat{\sigma}\hat{\Lambda}^2} (\nabla_{\tilde{x}}\tilde{u})^2 \, d^3\tilde{x} + \int_{\partial\hat{\mathscr{V}}} \frac{1}{\hat{\sigma}} \tilde{u}^2 \, d^2\tilde{\mathbf{x}}$$

$$+ \int_{\partial\mathscr{V}} \frac{N}{\hat{\sigma}\hat{\Lambda}^2} (1 - \tilde{u})^2 \, d^2\tilde{\mathbf{x}}$$

$$\geq \hat{\eta} \geq$$

$$- \int_{\hat{\mathscr{V}}} \frac{1}{\hat{\sigma}\hat{\Lambda}^2} \tilde{q} \cdot \tilde{q} \, d^3\tilde{x} - \int_{\partial\hat{\mathscr{V}}} \frac{1}{\hat{\sigma}\hat{\Lambda}^4} (\mathbf{n} \cdot \tilde{q})^2 \, d^2\tilde{x}$$

$$+ \int_{\partial\mathscr{V}} \frac{1}{\hat{\sigma}\hat{\Lambda}^2} \left[2\mathbf{n} \cdot \tilde{q} - N^{-1}(\mathbf{n} \cdot \tilde{q})^2 \right] d^2\tilde{x} \tag{4.8.6}$$

Note that the equality holds when $\tilde{u} = c/c_f$ and $\tilde{q} = L\nabla c/2c_f$, where c is the concentration distribution in the pellet.

Now assume $\mathscr{N}$ data points $(\hat{\Lambda}_i, N_i, \hat{\eta}_i)$ are given, then there must exist a concentration profile $\tilde{u}_i$ and a flux field $\tilde{q}_i$ corresponding to each data point. The functions

$$\tilde{u} = \sum_{i=1}^{\mathscr{N}} \alpha_i \tilde{u}_i \tag{4.8.7}$$

and

$$\tilde{q} = \sum_{i=1}^{\mathscr{N}} \beta_i \tilde{q}_i \tag{4.8.8}$$

where $\{\alpha_i\}$ and $\{\beta_i\}$ are two sets of positive numbers satisfying

$$\sum_{i=1}^{\mathscr{N}} \alpha_i = \sum_{i=1}^{\mathscr{N}} \beta_i = 1 \tag{4.8.9}$$

are admissible trial functions for bounding the effectiveness factor $\hat{\eta}$. Substituting these trial functions into the bounds (4.8.6), using the fact that the expected value of the squares is always greater than or equal to the square of the expected value, we obtain

$$\sum_{i=1}^{\mathscr{N}} \frac{\alpha_i}{\hat{\sigma}} \left\{ \int_{\mathscr{V}} \frac{1}{\hat{\Lambda}^2} (\nabla_{\tilde{x}} \tilde{u}_i)^2 \, d^3\tilde{x} + \int_{\partial\mathscr{V}} \tilde{u}_i^2 \, d^2\tilde{x} \right.$$

$$+ \int_{\partial\mathscr{V}} \frac{N}{\hat{\Lambda}^2} (1 - \tilde{u}_i)^2 \, d^2\tilde{x} \Bigg\}$$

$$\geq \hat{\eta} \geq$$

$$- \sum_{i=1}^{\mathscr{N}} \frac{\beta_i}{\hat{\sigma}} \left\{ \int_{\mathscr{V}} \frac{1}{\hat{\Lambda}^2} \tilde{q}_i \cdot \tilde{q}_i \, d^3\tilde{x} + \int_{\partial\mathscr{V}} \frac{1}{\hat{\Lambda}^4} (\boldsymbol{n} \cdot \tilde{q}_i)^2 \, d^2\tilde{x} \right.$$

$$- \int_{\partial\mathscr{V}} \frac{1}{\hat{\Lambda}^2} \left[2\boldsymbol{n} \cdot \tilde{q}_i - N^{-1}(\boldsymbol{n} \cdot \tilde{q}_i)^2 \right] d^2\tilde{x} \Bigg\} \tag{4.8.10}$$

At any one of the $\mathscr{N}$ data points $\hat{\Lambda}_i$, N_i, $\hat{\eta}_i$ the equality signs of (4.8.6) will hold when $\tilde{u}_i$ and $\tilde{q}_i$ are used in place of the trial functions, if now $\{\hat{a}_i\}$ and $\{\hat{b}_i\}$ are two sets of constants, we can rewrite (4.8.10) as

$$\sum_{i=1}^{\mathscr{N}} \frac{\alpha_i}{\hat{\sigma}} \left\{ \hat{a}_i \hat{\sigma} \hat{\eta}_i - \int_{\mathscr{V}} \left(\hat{a}_i - \frac{\hat{\Lambda}_i^2}{\hat{\Lambda}^2} \right) \frac{1}{\hat{\Lambda}_i^2} (\nabla_{\tilde{x}} \tilde{u}_i)^2 \, d^3\tilde{x} \right.$$

$$- \int_{\partial\mathscr{V}} (\hat{a}_i - 1) \tilde{u}_i^2 \, d^2\tilde{x}$$

$$- \int_{\partial\mathscr{V}} \left(\hat{a}_i - \frac{N}{\hat{\Lambda}^2} \frac{\hat{\Lambda}_i^2}{N_i} \right) \frac{N_i}{\hat{\Lambda}_i^2} (1 - \tilde{u}_i)^2 \, d^2\tilde{x} \Bigg\}$$

$$\geq \hat{\eta} \geq$$

$$\sum_{i=1}^{\mathscr{N}} \frac{\beta_i}{\hat{\sigma}} \left\{ \hat{b}_i \hat{\sigma} \hat{\eta}_i + \int_{\mathscr{V}} \left(\hat{b}_i - \frac{\hat{\Lambda}_i^2}{\hat{\Lambda}^2} \right) \frac{1}{\hat{\Lambda}_i^2} \tilde{q}_i \cdot \tilde{q}_i \, d^3\tilde{x} \right.$$

$$+ \int_{\partial\mathscr{V}} \left(\hat{b}_i - \frac{\hat{\Lambda}_i^4}{\hat{\Lambda}^4} \right) \frac{1}{\hat{\Lambda}_i^4} (\boldsymbol{n} \cdot \tilde{q}_i)^2 \, d^2\tilde{x}$$

$$- 2 \left(\hat{b}_i - \frac{\hat{\Lambda}_i^2}{\hat{\Lambda}^2} \right) \hat{\sigma} \hat{\eta}_i$$

$$+ \int_{\partial\mathscr{V}} \left(\hat{b}_i - \frac{\hat{\Lambda}_i^2 N_i}{\hat{\Lambda}^2 N} \right) \frac{1}{\hat{\Lambda}_i^2 N_i} (\boldsymbol{n} \cdot \tilde{q}_i)^2 \, d^2\tilde{x} \Bigg\} \tag{4.8.11}$$

Note that $\hat{\sigma}$ depends on the surface $\partial\hat{\mathcal{V}}$, hence it must remain the same at the various data points.

We further specify for the sets $\{\hat{a}_i\}$ and $\{\hat{b}_i\}$ that

$$\hat{a}_i = \max\left[\, 1,\, \frac{\hat{\Lambda}_i^2}{\hat{\Lambda}^2},\, \frac{N\hat{\Lambda}_i^2}{N_i\hat{\Lambda}^2} \,\right]$$

$$\hat{b}_i = \max\left[\, \frac{\hat{\Lambda}_i^2}{\hat{\Lambda}^2},\, \frac{\hat{\Lambda}_i^4}{\hat{\Lambda}^4},\, \frac{\hat{\Lambda}_i^2 N_i}{\hat{\Lambda}^2 N} \,\right] \tag{4.8.12}$$

to make the terms in curly brackets positive, then replacing these terms by something smaller only reinforces the inequality. On this basis, the volume integrals in both the upper and lower bounds can be deleted from (4.8.11). Applying the Schwartz inequality to either of the surface integrations ($\partial\mathcal{V}$ and $\partial\hat{\mathcal{V}}$) in (4.8.11), we have

$$\sum_{i=1}^{N} \alpha_i \left\{ \hat{a}_i\hat{\eta}_i - (\hat{a}_i - 1)\hat{\eta}_i^2 - \left(\hat{a}_i - \frac{N\hat{\Lambda}_i^2}{\hat{\Lambda}^2 N_i} \right)\left(\frac{\hat{\Lambda}_i^2}{N_i}\, \frac{\hat{s}L}{2} \right)\hat{\eta}_i^2 \right\}$$

$$\geq \hat{\eta} \geq$$

$$\sum_{i=1}^{N} \beta_i \left\{ \hat{b}_i\hat{\eta}_i + \left(\hat{b}_i - \frac{\hat{\Lambda}_i^4}{\hat{\Lambda}^4} \right)\hat{\eta}_i^2 - 2\left(\hat{b}_i - \frac{\hat{\Lambda}_i^2}{\hat{\Lambda}^2} \right)\hat{\eta}_i \right.$$

$$\left. + \left(\hat{b}_i - \frac{\hat{\Lambda}_i^2 N_i}{\hat{\Lambda}^2 N} \right)\left(\frac{\hat{\Lambda}_i^2}{N_i}\, \frac{\hat{s}L}{2} \right)\hat{\eta}_i^2 \right\} \tag{4.8.13}$$

where $\hat{s}$ is the active catalyst surface per unit total volume. Then choosing the sets $\{\alpha_i\}$ and $\{\beta_i\}$ so as to give the best possible bounds leads to

$$\min_i \left\{ \hat{a}_i\hat{\eta}_i - (\hat{a}_i - 1)\hat{\eta}_i^2 - \left(\hat{a}_i - \frac{N\hat{\Lambda}_i^2}{N_i\hat{\Lambda}^2} \right)\left(\frac{\hat{\Lambda}_i^2}{N_i}\, \frac{\hat{s}L}{2} \right)\hat{\eta}_i^2 \right\}$$

$$\geq \hat{\eta} \geq$$

$$\max_i \left\{ \hat{b}_i\hat{\eta}_i + \left(\hat{b}_i - \frac{\hat{\Lambda}_i^4}{\hat{\Lambda}^4} \right)\hat{\eta}_i^2 - 2\left(\hat{b}_i - \frac{\hat{\Lambda}_i^2}{\hat{\Lambda}^2} \right)\hat{\eta}_i \right.$$

$$\left. + \left(\hat{b}_i - \frac{\hat{\Lambda}_i^2 N_i}{\hat{\Lambda}^2 N} \right)\left(\frac{\hat{\Lambda}_i^2}{N_i}\, \frac{\hat{s}L}{2} \right)\hat{\eta}_i^2 \right\} \tag{4.8.14}$$

Qualitatively, this feasible region is the domain between two bounding surfaces which pinch together at each data point. Figure 4.8.1 shows the qualitative nature of this region if all the data is taken at constant N. If in taking data, a set of points lies far outside the region of feasibility constructed from the previous data points, then it or they are open to suspicion of their accuracy.

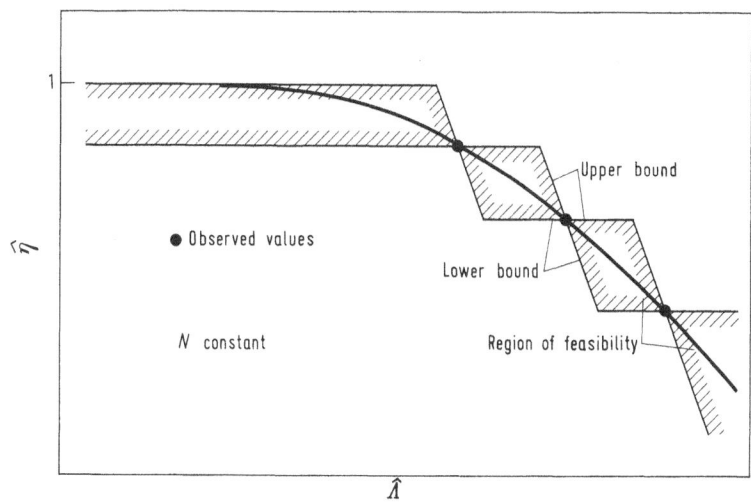

Fig. 4.8.1. Qualitative nature of the region of feasibility.

4.9. Non-monotonic Kinetics

When the reaction in a catalyst particle is exothermic the equations can be reduced to the form of equations (4.2.1) and (4.2.2), but the resulting rate function $r(c)$ is not necessarily monotonic. For example, for the first-order irreversible disappearance of a reactant whose concentration is $c(x)$ we have

$$\nabla \left(D_e \, \nabla c \right) = \left[A \exp - (E/RT) \right] c \qquad (4.9.1)$$

where $T = T(x)$ is the temperature at any point in $\mathscr{V}$. The temperature is itself subject to a similar differential equation obtained from an energy balance, namely

$$\nabla \cdot \left(k_e \nabla T \right) + \left(- \Delta H \right) \left[A \exp - (E/RT) \right] c = 0, \qquad (4.9.2)$$

where k_e is the effective thermal conductivity and ΔH the heat of reaction. If the Dirichlet problem with

$$c = c_f, \qquad T = T_f \quad \text{on } \partial \mathscr{V} \qquad (4.9.3)$$

is considered and D_e and k_e are constant, then

$$\nabla^2 \left[\left(- \Delta H \right) D_e c + k_e T \right] = 0 \quad \text{in } \mathscr{V} \qquad (4.9.4)$$

and

$$\left(- \Delta H \right) D_e c + k_e T = \left(- \Delta H \right) D_e c_f + k_e T_f \quad \text{on } \partial \mathscr{V}. \qquad (4.9.5)$$

But the only solution to equations (4.9.4) and (4.9.5) is that $(-\Delta H) D_e c + k_e T$ is everywhere constant and hence

$$T = T_f + \frac{(-\Delta H) D_e}{k_e}(c_f - c). \qquad (4.9.6)$$

By substituting the relation (4.9.6) in (4.9.1) we have indeed an equation of the same form as (4.2.1) but

$$r(c) = A\left[\exp - E/R\left\{ T_f + \frac{(-\Delta H) D_e}{k_e}(c_f - c)\right\} \right] c \qquad (4.9.7)$$

is not necessarily monotonic.

If however the temperature distribution were known so that equation (4.9.1) could be written

$$D_e \nabla^2 c = f(x) c \qquad (4.9.8)$$

then it would be immediately amenable to solution by variational methods since the solution of equations (4.9.8) and (4.9.2) would minimize

$$\mathscr{J}(u) = \left[\int_{\mathscr{V}} \{D_e \nabla u \cdot \nabla u + f(x) u^2\} d^3x \right] / V f^* c_f, \qquad (4.9.9)$$

where f^* denotes the value of $f(x)$ on the surface, namely

$$f^* = A \exp - E/R T_f. \qquad (4.9.10)$$

The trial functions u can be any piecewise twice differentiable functions in $\mathscr{V}$ with the value c_f on $\partial\mathscr{V}$.

This suggests an iterative scheme of calculation which would proceed as follows. Pick a trial temperature distribution $T_0(x)$ and calculate the corresponding $f_0(x)$. Let $\mathscr{J}_0(u)$ be the functional (4.9.9) with $f = f_0$ and $c_1(x)$ the function $u(x)$ which minimizes it. Calculate $T_1(x)$ from $c_1(x)$ by equation (4.9.6) and go through the same cycle. If this process converges a solution of equations (4.9.1)–(4.9.3) will be obtained. Rester (1972) has shown that this iterative process can give good numerical results. Its suitability to any particular problem would have to be judged on the merits of the individual case.

Bibliography

Abramowitz, M. and Stegun, I. A.: Handbook of Mathematical Functions, Washington, D. C.: U. S. Government Printing Office 1964.

Anderson, N., Arthurs, A. M. and Robinson, P. D.: Pairs of complementary variational principles. J. Inst. Math. Appl. **5**, 422 (1969).

Aris, R.: Shape factors for irregular particles, I. The steady state problem. Chem. Engng. Sci. **6**, 262 (1957).

────── The mathematical theory of diffusion and reaction in permeable catalysts, Oxford: Clarendon Press 1974.

Arthurs, A. M.: [1] Complementary variational principles in neutron diffusion theory. Proc. Roy. Soc. A **298**, 97 (1967).

────── [2] Extremum principles for a class of boundary value problems. Proc. Cambridge Phil. Soc. (Math., Phys., Sci.) **65**, 803 (1969).

────── [3] Complementary variational principles, Oxford: Clarendon Press 1970.

────── [4] Dual extremum principles and error bounds for a class of boundary value problems. J. Math. Anal. Applics. **41**, 781 (1973).

────── [5] Error bounds for a class of nonlinear problems in diffusion and reaction. J. Inst. Math. Appl. (to appear).

────── and Coles, C. W.: Error bounds on variational methods for nonlinear differential and integral equations. J. Inst. Math. Appl. **7**, 324 (1971).

Becker, M.: The principles and applications of variational methods, Cambridge, Mass.: M.I.T. Press 1964.

Beran, M. J.: Statistical continuum theories, New York: Interscience 1968.

Berman, A. S.: Variational upper bounds on Knudsen flow rates. J. Appl. Phys. **36**, 3356 (1965).

Biot, M. A.: Variational principles in heat transfer, Oxford: Clarendon Press 1970.

Brown, W. F.: [1] Solid mixture permittivities. J. Chem. Phys. **23**, 1514 (1955).

────── [2] Dielectric constants, permeabilities, and conductivities of random media. Trans. Soc. Rheol. **9:1**, 357 (1965).

Carman, P.: Flow of gases through porous media, London: Butterworths Scientific Publications, Ltd. 1956.

Courant, R. and Hilbert, D.: Methoden der mathematischen Physik, Bd.I., Berlin: Springer 1937. Translated as Methods of mathematical physics, vol. 1., New York: Interscience Pub. 1953.

Debye, P. and Bueche, A. M.: Scattering by an inhomogeneous solid. J. Appl. Phys. **20**, 518 (1949).

────── , Anderson, H. R. and Brumberger, H.: Scattering by an inhomogeneous solid II. The correlation function and its application. J. Appl. Phys. **28**, 679 (1957).

Derjaguin, B.: Measurement of the specific surface of porous and disperse bodies by their resistance to the flow of rarefied gases. Compt. Rend. Acad. Sci., URSS **53**, 623 (1946).

Donnelly, R. J., Herman R. and Prigogine, I.: (eds.) Non-equilibrium thermodynamics, variational techniques and stability, Chicago: Univ. of Chicago Press 1966.

Finlayson, B. A., Scriven, L. E.: Galerkin's method and the local potential, in Donnelly, Herman and Prigogine 1966.

Funk, P.: Variationsrechnung und ihre Anwendung in Physik und Technik, Berlin/
    Göttingen/Heidelberg: Springer 1962.
Gould, S. H.: Variational methods for eigenvalue problems, (2nd. ed.), Toronto: Univ.
    of Toronto Press 1966.
Gunn, D. J.: Diffusion and chemical reaction in catalysis and absorption. Chem. Engng.
    Sci. **22**, 1439 (1967).
Ham, F. S.: [1] Theory of diffusion limited precipitation. J. Phys. Chem. Solids. **6**,
    335 (1958).
——— [2] Diffusion-limited growth of precipitate particles. J. Appl. Phys. **30**, 1518 (1959).
——— [3] Stress-assisted precipitation on dislocations. J. Appl. Phys. **30**, 915 (1959).
Hashin, Z. and Shtrikman, S.: Variational approach to the theory of effective magnetic
    permeability of multiphase material. J. Appl. Phys. **33**, 3125 (1962).
Hill, R.: New derivations of some elastic extremum principles. Progress in Applied
    Mechanics, New York: The Prager Anniv. Vol., Macmillan Co. 1963.
Jackson, J. L. and Coriell, S. R.: Diffusion constant in a polyelectrolyte solution. J. Chem.
    Phys. **38**, 959 (1963).
Jüttner, F.: Reaktionskinetik und Diffusion. Z. phys. Chem. **65,** 595 (1909).
Keller, J. B.: Extremum principles for irreversible processes. J. Math. Phys. **11**, 2919 (1970).
Lanczos, C.: The variational principles of mechanics, (3rd ed.), Toronto: Univ. of
    Toronto Press 1966.
Lassettre, E.: Influence of specular reflection on the permeability of porous media, U. S.
    Atomic Energy Commission Report K-1258, 1956.
Lauwerier, H. A.: Calculus of variations in mathematical physics, Amsterdam: Mathe-
    matical Centrum 1966.
Luss, D. and Amundson, N. R.: On a conjecture of Aris: proof and remarks. A.I.Ch.E.
    Journal **13**, 759 (1967).
deMarcus, W. C.: Variational upper bounds on Knudsen flow rates-capillary. Advances
    in applied mechanics suppl. 1 Rarefied gas dynamics, New York: Academic Press
    Inc. 1961.
Noble, B.: Complementary variational principles for boundary value problems 1: Basic
    principles with an application to ordinary differential equations, Univ. of Wisc. Math.
    Res. Center Rep. No. 473, 1964.
——— and Sewell, M. J.: On dual extremum principles in applied mathematics. J. Inst.
    Math. Appl. **9**, 123 (1972).
Noyes, R. M.: Effects of diffusion rates on chemical kinetics. Progress in reaction kinetics,
    (vol. 1), London: Pergamon Press Ltd. 1961.
Polya, G. and Szego, G.: Isoperimetric inequalities in mathematical physics, Princeton,
    N.J.: Princeton Univ. Press 1951.
Prager, S.: [1] Diffusion and viscous flow in concentrated suspensions. Physica. **29**, 129
    (1963).
——— [2] Interphase transfer in stationary two-phase media. Chem. Engng. Sci. **18**, 227
    (1963).
——— [3] Improved variational bounds on some bulk properties of a two-phase random
    medium. J. Chem. Phys. **50**, 4305 (1969).
Reck, R. A. and Prager, S.: Diffusion-controlled quenching at higher quencher concen-
    trations. J. Chem. Phys. **42**, 3027 (1965).
——— and Reck, G. P.: [1] Theory of diffusion-controlled reactions in porous media.
    J. Chem. Phys. **49**, 701 (1968).
———, ——— [2] Theory of diffusion-controlled reactions in porous media. J. Chem.
    Phys. **49**, 3618 (1968).
Rester, S.: Topics in diffusion and reaction, Ph.D. Dissertation, University of Minnesota
    1972, University Microfilms Ann Arbor. Michigan.
——— and Aris, R.: Communications on the theory of diffusion and reaction VIII.

Variational bounds on the effectiveness factor. Chem. Engng. Sci. **27**, 347 (1972).

Rotne, J. and Prager, S.: Treatment of hydrodynamic interaction in polymers. J. Chem. Phys. **50**, 4831 (1969).

Satterfield, C. N.: Mass transfer in heterogeneous catalysis, Cambridge, Massachusetts: M.I.T. Press 1970.

―――― and Roberts, G. W.: Effectiveness factor for porous catalysts. Langmuir-Hinshelwood kinetic expressions. Ind. Eng. Chem. (Fundamentals) **4**, 288 (1965).

Schechter, R. S.: The variational method in engineering, New York: McGraw-Hill 1967.

Smoluchowski, M.: [1] Über Brownsche Molekularbewegung unter Einwirkung äußerer Kräfte und deren Zusammenhang mit der verallgemeinerten Diffusionsgleichung. Ann. Phys. **48**, 1103 (1915).

―――― [2] Versuch einer mathematischen Theorie der Koagulationskinetik kolloiden Lösungen. Z. phys. Chem. **92**, 129 (1917).

Strieder, W. and Prager, S.: [1] Upper and lower bounds on Knudsen flow rates. J. Math. Phys. **3**, 514 (1967).

――――, ―――― [2] Knudsen flow through a porous medium. J. Phys. Fluids **11**, 2544 (1968).

Strieder, W.: [1] Knudsen flow and chemical reaction in a porous catalyst. J. Chem. Phys. **51**, 566 (1969).

―――― [2] Knudsen diffusion and chemical reaction in a random bed of solid spheres. J. Chem. Phys. **52**, 5204 (1970).

―――― [3] Gaseous self-diffusion through a porous medium. J. Chem. Phys. **54**, 4050 (1971).

―――― and Aris, R.: Variational bounds for problems in diffusion and reaction. J. Inst. Math. Appl. **8**, 328 (1972).

―――― and Kohn, J.: Simpler solutions of some well-known boundary value problems. Ind. Eng. Chem. (Fundamentals) **11**, 593 (1972).

Synge, J. L.: The hypercircle in mathematical physics, London: Cambridge Univ. Press 1957.

Thiele, E. W.: The effect of grain size on catalyst performance. Amer. Sci. **55**, 176 (1967).

Weissberg, H. L.: Effective diffusion coefficient in porous media. J. Appl. Phys. **34**, 2636 (1963).

―――― and Prager, S.: Viscous flow through porous media II. Approximate three point correlation function. Phys. Fluids **5**, 1390 (1962).

Woodbury, G. W. and Prager, S.: Motion in many particle systems. I. Forced inter-diffusion of two species. J. Chem. Phys. **38**, 1446. JACS **86**, 3417 (1963).

Young, L. C.: Lectures on the calculus of variations and optimal control theory, Philadelphia: W. B. Saunders 1969.

Notation

a	sphere radius	B^*	excited species in the quenching problem
a_i	$i = 1, 2, \ldots$ stoichiometric coefficients in (4.4.1)	$c(x)$	solute concentration at x
a_α, a_β	various sphere radii introduced in (2.6.17)	$\tilde{c}$	dimensionless concentration defined by (4.4.67)
$a_p(t)$	function of time introduced in (3.2.10)	$\bar{c}$	trial function defined for a spherical catalyst pellet to satisfy inequality (4.5.7)
$\hat{a}_i$	$i = 1, 2, \ldots$ set of constants introduced in (4.8.11)	$\langle c \rangle$	volume average of excited species concentration defined by (3.5.3)
A	constant vector in the direction of the average concentration gradient in (2.6.3a)	c_0, c_L	concentrations at the ends $x = 0$ and $x = L$ of a long slab
A	magnitude of A in (2.6.4), also used as reactant species in (4.4.3)	c_f	reactant concentration in the bulk phase far from the catalyst pellet
A_p	normalizing constant for the eigenfunction (3.3.1)	c_p	concentration at sphere center in (2.5.5a)
A_i	reactants in (4.4.1)	c_s	saturated concentration of solute
A_q	quencher species		
$[A_q]$	quencher concentration	$c_0 + c_s$	initial concentration of solute defined by (3.2.3)
$[A_q] f^{(1)}(\boldsymbol{\rho}_1)$	mean concentration of A_q at $x + \boldsymbol{\rho}_1$ if x is in $\mathcal{V}$	c_T	total concentration in a binary mixture of A and B
$[A_q]^2 f^{(2)}(\boldsymbol{\rho}_1, \boldsymbol{\rho}_2) \, d^3\boldsymbol{\rho}_1 \, d^3\boldsymbol{\rho}_2$	mean number of quencher pairs with two different quencher particles located respectively in the volume elements $d^3\boldsymbol{\rho}_1$ and $d^3\boldsymbol{\rho}_2$ at $\boldsymbol{\rho}_1$ and at $\boldsymbol{\rho}_2$ relative to a randomly selected point x in $\mathcal{V}$.	$c_v(t)$	defined in (3.3.9)
		$c_A, c_{A_i}, c_B, c_{B_i}$	$i = 1, 2, \ldots$ reactant A, A_i and product B, B_i concentrations in (4.4.2)
		$c_{A_{if}}, c_{B_{if}}$	$i = 1, 2, \ldots$ concentration of species A_i and B_i far from the catalyst pellet in (4.4.8)
$\mathcal{A}(q, u)$	function of q and u on $\partial \mathcal{V}$, introduced in (1.6.2)	$d^3x, d^3\rho$	infinitesmal elements of volume
$\mathcal{A}_u, \mathcal{A}_q, \mathcal{A}_{uu}, \mathcal{A}_{qu}, \mathcal{A}_{qq}$	vector and scalar derivatives of $\mathcal{A}$ introduced in equations (1.6.5) through (1.6.9)	d^2x	infinitesmal element of surface
		$d^2\boldsymbol{n}, d^2\boldsymbol{n}'$	elements of solid angle
b	parameter in (4.3.10)	$d^3\tilde{x}$	dimensionless volume element introduced in (4.8.6)
b_i	$i = 1, 2, \ldots$ stoichiometric coefficient in (4.4.1)		
$\hat{b}_i$	$i = 1, 2, \ldots$ set of constants introduced in (4.8.11)	$d^2\tilde{x}$	dimensionless element of surface area, defined by (4.8.5)
B	product species in (4.4.3)	$d\theta/dn$	magnitude of the concentration ($\theta = c$) gradient in an
B_i	$i = 1, 2, \ldots$ product species in (4.4.1)		

	arbitrary catalyst pellet, defined by (4.5.11)
$D, D(x), D_0$	local diffusion coefficients. D also used in place of D_e for the effective diffusion coefficient in section 4.5
D_e	effective diffusion coefficient
D_r, D_a, D_b	diffusion coefficients for the various phases of section 2.5
D_{eA_i}, D_{eB_i}	effective diffusion coefficients of species A_i and B_i
E	defined by (4.4.22)
$f(x)$	arbitrary function of position
$f_p(x)$	eigenfunction for (4.4.57)
$F(u)$	defined by (4.2.8)
$F_a(\rho)$	defined by (2.6.3b)
$F_G(\theta)$	defined by (1.3.14)
$\hat{F}(u)$	defined by (4.6.10)
$F_{a_\alpha}(\rho_i), F_{a_\beta}(\rho_i)$	defined by (2.6.18)
$\mathscr{F}(q, u)$	variational functional defined by (1.6.2)
$g(x)$	stochastic function of pore structure defined by (1.3.2)
$G(\rho, \rho')$	$(G(\rho, \rho') = \langle g(x)g(x+\rho)g(x+\rho')\rangle)$ three-point correlation
G_θ	magnitude of the concentration gradient, defined by (4.5.11)
$\mathscr{G}(q)$	variational functional (1.6.18) also used as variational lower bound functional for a catalyst pellet (4.2.15)
$\mathscr{G}_{max}$	maximum value of the variational functional $\mathscr{G}(q)$
$\hat{\mathscr{G}}(q)$	variational lower bound functional for a molecular sieve catalyst, defined by (4.6.17)
$\hat{\mathscr{G}}_{max}$	maximum value of the variational functional $\hat{\mathscr{G}}(q)$
h	length of a cylinder
$h(\rho)$	introduced as part of trial function (2.4.2)
$h_0(\rho)$	value of $h(\rho)$ that minimizes variational functional
$h_1(\rho)$	variation of the trial $h(\rho)$ about $h_0(\rho)$
$\tilde{h}, \tilde{h}_i$	$i = 0, 1$ defined by (2.4.20)
$h_\sigma(\rho, n, n')\,d^3\rho\,d^2n\,d^2n'$	probability that two points on the pore wall which can see one another have a relative position vector ρ lying in $d^3\rho$ and that the unit normal n to the pore wall

	lies within d^2n at the first point and within d^2n' about n' at the second
h_z	defined by (4.4.42)
$h^*(x)$	defined in (3.7.12) and (3.7.13), an improved form of this trial function is defined by (3.8.7)
$H(\rho)$	function of ρ only defined in (2.4.7)
H_s	defined by (4.5.17)
$H^*(\varepsilon)$	trial function defined by (3.6.12)
$\mathscr{H}(x, q, u)$	function of x, q, and u defined in $\mathscr{V}$ introduced in (1.6.2)
$\mathscr{H}_u, \mathscr{H}_q, \mathscr{H}_{uu}, \mathscr{H}_{qu}, \mathscr{H}_{qq}$	vector and scalar derivatives of $\mathscr{H}$ introduced in (1.6.5) through (1.6.9)
i_x	unit vector across a slab in the positive x-direction
$I(\nabla \cdot q)$	defined by (4.2.16)
$I_0(\alpha_p), I_1(\alpha_p)$	zeroth and first-order modified Bessel functions
$\hat{I}(n \cdot q)$	defined by (4.6.18)
$j(x)$	flux vector
J	mean flux, net rate at which molecules pass through a unit total cross-section of a slab of suspended solid
J	magnitude of the mean flux J
$\mathscr{J}(u)$	variational functional defined by (1.6.14), also used as a variational upper bound functional for a catalyst pellet (4.2.8)
$\mathscr{J}_{min}$	minimum value of the variational functional $\mathscr{J}(u)$
$\hat{\mathscr{J}}(u)$	variational upper bound functional for a molecular sieve catalyst, defined by (4.6.9)
$\hat{\mathscr{J}}_{min}$	minimum of the variational functional $\hat{\mathscr{J}}(u)$
k, k_r	first-order reaction rate constant per unit volume introduced in (4.5.2) and (4.3.1)
k_B	Boltzmann's constant
k_c	coefficient for external mass transfer
k_e	effective first-order reaction rate constant, defined by (3.5.4)
k	first-order surface reaction rate constant
k_r'	rate constant in the numerator of the Langmuir-Hinshelwood rate law (4.4.9)

K_Φ	defined by (2.4.14)
K	constant in the denominator of the Langmuir-Hinshelwood kinetic rate law (4.4.9)
$K(u)$	variational functional defined by (3.4.1)
$K(x, x')\, d^2x'$	probability that a molecule emitted from the pore wall at x will make its next wall collision within d^2x' at x'
K_{A_i}	adsorption coefficient for species A_i
l	radius of a circle passing through all the vertices of a triangle with sides ρ, ρ', and $(\rho' - \rho)$ defined by (1.3.13)
l_1, l_2, l_3	largest, second largest, and smallest of the sides of a triangle with sides ρ, ρ', and $(\rho' - \rho)$, introduced in (1.3.12)
$\dot{l}_z$	defined by (4.4.45)
$L/2$	ratio of volume of $\mathcal{V}$ to external surface area of $\partial\mathcal{V}$, for a slab this is one-half the thickness
$\mathcal{L}$	defined by (4.4.28)
m	adjustable constant in (2.6.3a)
m_a	$(m_a = 4\Phi/s)$ average pore diameter
m_α, m_β	adjustable constant in (2.6.18)
M	defined by (4.2.28)
$\mathcal{M}$	sphere of volume V
n	density of sphere centers in a random bed of solid spheres
$n, n', n(x)$	unit normals to a surface
n_α, n_β	densities of sphere centers for the radii $a_\alpha, a_\beta, \ldots$ respectively introduced in (2.6.17)
N	mass Biot number defined by (4.2.7)
N_a	number of sphere centers in a volume V introduced in (1.3.7)
N_i	mass Biot number at one of $i = 1, 2, 3, \ldots, \mathcal{N}$ data points
N_Φ	modified mass Biot number, defined by (4.7.12)
$\mathcal{N}$	number of known data points for a molecular sieve catalyst pellet
$P(\varepsilon)\, d\varepsilon$	probability that a point chosen at random in the void region lies at a distance between ε and $\varepsilon + d\varepsilon$ from the closest point on the interface
P_0, P_L	pressure at the two ends of a slab $x = 0$ and $x = L$ respectively
P_s	probability of finding at least one sphere center within a shell of radii $(\varepsilon + a)$ to $(\varepsilon + a) + d\varepsilon$
P_v	probability that the volume v in a random bed of spheres contains no sphere centers
$\bar{P}(\theta)$	surface average of the concentration gradient magnitude, defined by (4.5.13)
$\mathscr{P}$	probability that two points chosen at random on the surfaces of two different spheres are exposed and can see one another, given by (1.4.3)
$q(x)$	trial flux vector field
$q(\rho, n, n')$	arbitrary function defined in (2.9.3)
q_μ	defined by (4.4.31)
$q^*(x_i - x)$	defined by (3.7.13)
$q_0(x)$	trial flux that maximizes the variational functional
$q_1(x)$	variation of the trial flux q about q_0
$\tilde{q}$	dimensionless trial flux, defined by (4.8.3)
$\tilde{q}_i$	$i = 1, 2, \ldots, \mathcal{N}$ dimensionless flux distribution for one of the known data points
δq_δ	trial flux in excess of maximizing value from section 1.6
Q, Q_+, Q_-	defined by (4.4.23) and (4.4.24)
$\mathcal{Q}$	defined by (4.4.34)
r	radius of a cylinder
$r(c)$	reaction rate per unit volume of a homogeneous catalyst pellet
r_a, r_b	radii of the smaller and larger of two concentric circles
r_f	reaction rate $r(c)$ evaluated at c_f
r_s	radius of the sphere with the same volume as the cubic cell of section 3.2
r_σ	radial vector for spherical coordinates
r'	first derivative of the reaction rate $r(c)$

r^{-1}	inverse of the reaction rate function $r(c)$		$u_1(x)$	variation of the trial concentration u about u_0
$\hat{r}(c)$	reaction rate on the catalyst internal active surface per unit of surface		$u'(x)$	trial fluctuation about the mean concentration, defined in (2.3.7)
$\hat{r}_f$	reaction rate $\hat{r}(c)$ evaluated at c_f		$\tilde{u}$	dimensionless concentration, defined by (4.8.2)
$\hat{r}'$	first derivative of the reaction rate $\hat{r}(c)$		$\tilde{u}_i$	$i = 1, 2, \ldots \mathcal{N}$ dimensionless concentration distribution for one of $\mathcal{N}$ known data points
$\hat{r}^{-1}$	inverse of the surface reaction rate function $\hat{r}(c)$		$\delta u_\delta(x)$	trial scalar field in excess of minimizing function, used in section 1.6
$(r^{-1})'$	derivative of the inverse function r^{-1}		v	finite volume from which sphere centers are excluded, introduced in (1.3.7)
$(\hat{r}^{-1})'$	derivative of the inverse function $\hat{r}^{-1}$		$\bar{v}$	mean molecular speed
R	radius of a sphere whose volume is fixed equal to that of the cylinder in section 4.5		V	total volume
			$V(\theta)$	volume enclosed by a surface Σ_θ of constant concentration $c = \theta$ within a catalyst pellet, introduced in (4.5.10)
$R(c)$	defined by (4.2.29)			
$\hat{R}(c)$	defined by (4.6.36)			
R_a	constant defined by (3.4.15)		$V'(\theta)$	derivative of $V(\theta)$
R^*	radius of a spherical region drawn concentric with each spherical quencher molecule, defined in (3.7.13)		V_a	large volume excluding a sphere of radius a at the origin
			V_b	volume of a composite sphere of radius r_b, introduced in (2.5.1)
$\mathscr{R}$	region in $\mathscr{V}$			
s	pore wall surface per unit total volume		V_s	volume of a sphere of radius a
$\hat{s}$	catalytically active surface area per unit total volume of molecular sieve catalyst pellet		$\mathscr{V}$	arbitrary region of volume V
			$\hat{\mathscr{V}}$	subvolume of the total volume $\mathscr{V}$
$\mathscr{S}$	void-solid interface		$V_\Omega(\boldsymbol{\rho})$	volume of the region Ω of all points lying within a distance a of either end of the vector $\boldsymbol{\rho}$, given by (1.3.10)
$S(\boldsymbol{\rho})$	$(S(\boldsymbol{\rho}) = \langle g(\boldsymbol{x}) g(\boldsymbol{x} + \boldsymbol{\rho}) \rangle)$, two-point correlation			
$\mathscr{S}_c$	outer wall of the cubic cell, introduced in (3.2.4)		$V_{\Omega'}(\boldsymbol{\rho}, \boldsymbol{\rho}')$	volume of the region Ω' of all points lying at a distance a or less from one or more of the three vertices of the triangle with sides $\boldsymbol{\rho}$, $\boldsymbol{\rho}'$, and $(\boldsymbol{\rho}' - \boldsymbol{\rho})$, given by (1.3.12)
$\mathscr{S}_0, \mathscr{S}_L$	intersection of planes $x = 0$ and $x = L$ with the void regions of a slab			
$S_{\partial \mathscr{V}}$	outside surface area of $\mathscr{V}$, area of surface $\partial \mathscr{V}$			
$\hat{S}$	total internal catalytically active area within a molecular sieve catalyst pellet, area of $\partial \hat{\mathscr{V}}$		W, W_+, W_-	defined by (4.4.37)
			x	length coordinate across the slab
$\mathscr{S}$	arbitrary surface in $\mathscr{V}$		$\boldsymbol{x}, \boldsymbol{x}'$	position vectors
t	time		$\boldsymbol{x}_i$	position of the ith sphere center
T	temperature		$\tilde{\boldsymbol{x}}, \tilde{x}$	dimensionless position vector and length coordinates, defined by (4.8.4)
$\mathscr{T}$	defined by (4.4.35)			
$u(x)$	trial concentration scalar field		y	variable of integration in (3.8.3)
$u_0(x)$	trial concentration that minimizes the variational functional		Y	defined by (4.4.36)
			Z	defined by (4.4.40)

0, 1, 2 subscripts on variational functionals refer to forms zeroth, first, and second order in the variation of the trial function

α	magnitude of the vector $\boldsymbol{x}$	∂V_b	external surface of a composite sphere of radius r_b introduced in (2.5.2)
$\boldsymbol{\alpha}$	mean concentration gradient $\langle \nabla c \rangle$, also $(\psi_L - \psi_0)\boldsymbol{i_x}/L$ in (2.8.4)		
α_a, α_b	arbitrary constants in (2.5.6) and (2.5.7)	$\partial \hat{\mathscr{V}}$	external surface of the sub-volume $\hat{\mathscr{V}}$
α_i	set of $\mathscr{N}$ positive numbers $\sum_{i=1}^{\mathscr{N}} \alpha_i = 1$ introduced in (4.8.7)	$\Delta \mathscr{V}(\theta)$	element of volume between two surfaces of constant concentration $c = \theta$ and $c = \theta + d\theta$ in an arbitrary catalyst pellet, introduced in (4.5.12)
β	parameter in trial concentration (4.3.10)		
$\beta(\boldsymbol{x})$	Langrangian multiplier introduced in (3.7.3)	$\varepsilon(\boldsymbol{x})$	minimum distance from point $\boldsymbol{x}$ in $\hat{\mathscr{V}}$ to the reactive interface $\partial \hat{\mathscr{V}}$, introduced in (1.5.1) and (3.6.12)
β_i	set of $\mathscr{N}$ positive numbers $\sum_{i=1}^{\mathscr{N}} \beta_i = 1$ introduced in (4.8.8)		
$\beta_{\mathscr{J}_{\min}}, \beta_{\mathscr{G}_{\max}}$	values of β that respectively minimize $\mathscr{J}$ and maximize $\mathscr{G}$, listed in Table 4.4.3	ε_b	arbitrary constant in (2.5.6)
		ζ	variable of integration
		ζ_p	variable introduced in (3.7.18)
γ	cube root of diameter to height ratio for a cylinder and as a constant in (1.7.1)	η	effectiveness factor for a catalyst pellet
γ_c	scalar function defined by (4.2.20)	η_c	effectiveness factor for a finite cylinder
γ_e	equilibrium distribution coefficient for the solute between the solid and void phases at the pore walls	$\hat{\eta}$	effectiveness factor for a molecular sieve catalyst pellet
		$\hat{\eta}_i$	$i = 1, 2, \dots \mathscr{N}$ effectiveness factor at any one of $\mathscr{N}$ known data points
γ_q	mean number of quencher particles in a sphere of radius a, defined by (3.8.3)	θ	parameter used in spherical symmetrization to establish inequality (4.5.7)
γ_0	the extremum value of the variational functional $K(u)$, defined by (3.4.3)	θ_i	interior angles of a triangle with sides $\boldsymbol{\rho}$, $\boldsymbol{\rho'}$, and $(\boldsymbol{\rho'} - \boldsymbol{\rho})$, introduced in (1.3.12)
γ_1	positive constant in (4.4.47) and (4.4.48)	θ_m	parameter with a value between zero and one
$\gamma_{\min}$	defined by (4.4.59)	θ_σ	polar angle for spherical coordinates
$\tilde{\gamma}_{\min}$	defined by (4.4.65)	$\bar{\theta}$	radial coordinate in a spherical catalyst pellet selected so that the volume enclosed by the trial function $\bar{c}(\bar{\theta})$ in the sphere and the volume enclosed by the concentration value $c = \theta$ in an arbitrary catalyst pellet are equal, defined by (4.5.10)
$\Gamma[\xi]$	variational functional for Knudsen diffusion, defined by (2.8.1)		
$\Gamma_{mm}, \Gamma_{mf}, \Gamma_{ff}$	mean and fluctuating parts of $\Gamma[\xi]$ defined by equations (2.8.6) through (2.8.8)		
δ	parameter introduced in (1.6.3)	$\Theta(\boldsymbol{q})$	scalar field defined by (1.6.17)
$\delta(\boldsymbol{\rho}_i)$	Dirac delta function	κ	$[\kappa = c_f K]$ dimensionless number for Langmuir-Hinshelwood kinetics, defined by (4.4.18)
$\delta_{pp'}$	Kronecker delta		
$\partial \mathscr{V}$	external surface of total volume $\mathscr{V}$		

λ	adjustable parameter	τ	measured tortuosity factors, see Figure (2.6.1)
λ_{opt}	optimum value of λ		
λ_p^2	eigenvalue for (3.2.7)	τ_1	tortuosity bound for a bed of spheres, given by (2.6.14)
Λ	Thiele modulus defined by (4.2.6)		
$\Lambda_{\hat{\Phi}}$	modified Thiele modulus, defined by (4.7.12)	τ_m	tortuosity bound for any isotropic suspension, given by (2.6.16)
$\hat{\Lambda}$	Thiele modulus for a molecular sieve catalyst, defined by (4.6.6)	τ_p	decay time, defined by (3.2.14)
		$\Upsilon'(u)$	upper bound variational functional for diffusion controlled quenching, defined by (3.6.1)
$\hat{\Lambda}_i$	$i = 1, 2, \ldots \mathcal{N}$ Thiele modulus at one of $\mathcal{N}$ known data points		
		$\Upsilon''(u)$	upper bound variational functional with added constraint, defined by (3.6.4)
μ	defined by (4.4.26)		
μ_1	non-negative constant in (4.4.47) and (4.4.48)	Υ_{min}	minimum value of $\Upsilon(u)$, given by (3.6.10)
ν	variable used in (4.2.22) and (4.6.28)	ϕ	defined by (4.5.5)
		ϕ_c	defined by (4.5.2)
ξ	parameter in (4.3.6) and (4.4.19)	ϕ_σ	azimuth angle for spherical coordinates
$\xi(x)$	trial function for Knudsen diffusion, introduced in (2.8.1)		
$\xi_0(x)$	trial function that minimizes the variational functional $\Gamma[\xi]$	Φ	void fraction, a ratio of void volume to total volume
		$\hat{\Phi}$	inert volume to total volume in a molecular sieve catalyst
$\xi_1(x)$	variation of the trial function $\xi(x)$ about $\xi_0(x)$	χ	dimensionless trial concentration at slab edge, introduced in (4.3.6), (4.3.10), (4.4.19), and (4.4.32)
$\Xi(q)$	lower bound variational functional for diffusion controlled quenching, defined by (3.7.1)		
Π	permeability, mean molecular flux per unit pressure gradient	$\chi_{\mathcal{I}_{\text{min}}}, \chi_{\mathcal{G}_{\text{max}}}$	values of χ that respectively minimize $\mathcal{I}$ and maximize $\mathcal{G}$, see Table (4.4.3)
ρ, ρ', ρ''	relative position vectors	$\psi(x)$	rate at which molecules are reemitted from a unit element of pore wall surface located at x, introduced in (2.7.3)
ρ_i	$[\rho_i = (x_i - x)]$, relative position vector		
σ	mean pore surface area which can be seen from a typical point on the void-solid interface, given by (1.4.5)	ψ_0, ψ_L	defined by (2.7.2)
		ψ_m, ψ_f	mean and fluctuating parts of the trial function for Knudsen diffusion, introduced in (2.8.5)
σ^*	uniform generation rate of excited species B^* introduced in (3.5.1)	$\psi_p(x)$	eigenfunctions for (3.2.7)
		$\psi_{2/3}(\gamma_q)$	incomplete Gamma function
$\hat{\sigma}$	dimensionless surface area for catalytically active surface, defined by (4.6.7)	ω	Lagrangian multiplier in (3.6.4)
		$\omega(x)$	Lagrangian multiplier in (4.6.17)
$\Sigma(u)$	upper bound variational functional for diffusion, defined by (2.3.1)	$\bar{\omega}_p^2$	dimensionless eigenvalue defined by (4.4.64)
		ω_p^2	eigenvalue for (4.4.57)
Σ_b	dissipation integral for a composite sphere, defined by (2.5.1)	Ω	a region of all points lying within a distance a of either end of the vector ρ
Σ_θ	surface of constant concentration $c = \theta$, introduced in (4.5.12)	Ω'	a region of all points lying at a distance a or less from one or

more of the three vertices of the triangle with sides ρ, ρ', and $(\rho' - \rho)$

$\langle \ldots \rangle$ volume average defined by (1.3.1)

$\langle \ldots \rangle_{\mathscr{S}}$ surface average over the void-solid interface, defined by (2.9.2)

∇ gradient operator

$\nabla_{\tilde{x}}$ gradient operator in dimensionless coordinates $\tilde{x}$

$\|u - c\|_{\mathscr{V}}$ defined by (4.4.61)

Author Index

Abramowitz, M.   56, 97
Amundson, N. R.   79, 83, 98
Anderson, H. R.   9, 97
Anderson, N.   97
Aris, R.   59, 65, 79, 85, 91, 97, 98, 99
Arthurs, A. M.   13, 16, 61, 71, 75, 97

Becker, M.   97
Beran, M. J.   97
Berman, A. S.   97
Biot, M. A.   1, 2, 97
Brown, W. F.   97
Brumberger, H.   9, 97
Bueche, A. M.   6, 97

Carman, P.   33, 34, 41, 97
Coles, C. W.   75, 97
Coriell, S. R.   98
Courant, R.   1, 97

Debye, P.   5, 6, 9, 97
Derjaguin, B.   41, 97
Donnelly, R. J.   1, 97

Finlayson, B. A.   1, 97
Funk, P.   1, 97

Gould, S. H.   1, 98
Gunn, D. J.   79, 98

Ham, F. S.   43, 98
Hashin, Z.   98
Herman, R.   1, 97
Hilbert, D.   1, 97
Hill, R.   98

Jackson, J. L.   98
Jüttner, F.   59, 98

Keller, J. B.   98
Kohn, J.   79, 99

Lanczos, C.   1, 98
Lassettre, E.   41, 98
Lauwerier, H. A.   1, 98
Luss, D.   79, 83, 98

deMarcus, W. C.   2, 35, 36, 98

Noble, B.   13, 16, 98
Noyes, R. M.   98

Ockham, William of   1

Polya, G.   80, 82, 98

Prager, S.   8, 11, 30, 49, 91, 98, 99
Prigogine, I.   1, 97

Reck, G. P.   11, 49, 98
Reck, R. A.   11, 49, 98
Rester, S.   65, 85, 91, 96, 98
Roberts, G. W.   69, 99
Robinson, P. D.   97
Rotne, J.   98

Satterfield, C. N.   69, 99
Schechter, R. S.   99
Scriven, L. E.   1, 97
Sewell, M. J.   13, 98
Shtrikman, S.   98
Smoluchowski, M.   56, 99
Stegun, I. A.   56, 97
Strieder, W.   8, 79, 99
Synge, J. L.   99
Szego, G.   80, 82, 98

Thiele, E. W.   59, 99

Weissberg, H. L.   5, 6, 8, 30, 99
Woodbury, G. W.   99

Young, L. C.   1, 99

Subject Index

Springer Tracts in Natural Philosophy